MICROBIAL BIOMASS PROTEINS

MICROBIAL BIOMASS PROTEINS

Edited by

MURRAY MOO-YOUNG
University of Waterloo, Canada

and

KENNETH F. GREGORY
University of Guelph, Canada

ELSEVIER APPLIED SCIENCE
LONDON and NEW YORK

ELSEVIER APPLIED SCIENCE PUBLISHERS LTD
Crown House, Linton Road, Barking, Essex IG11 8JU, England

Sole Distributor in the USA and Canada
ELSEVIER SCIENCE PUBLISHING CO., INC.
52 Vanderbilt Avenue, New York, NY 10017, USA

WITH 77 TABLES AND 31 ILLUSTRATIONS

British Library Cataloguing in Publication Data

Microbial biomass proteins.
1. Food—Protein content
I. Moo-Young, Murray
II. Gregory, Kenneth F.
641.1′2 TX553.P7

ISBN 1-85166-085-2

Library of Congress CIP data applied for

The selection and presentation of material and the opinions expressed are the sole responsibility of the author(s) concerned

Printed in Great Britain by Galliard (Printers) Ltd, Great Yarmouth

PREFACE

This publication on 'microbial biomass proteins' (MBP) is the result of significant renewed interest in the subject matter. The title is chosen in an attempt to redress the historical anomaly whereby the term 'single cell proteins' (SCP), originally proposed in 1967 for these types of products, is no longer valid. Recent events have shown that the mass cultivation of multi-cellular fungi in addition to single-celled bacteria and yeasts is of commercial interest for the protein content of these microorganisms as animal and human food ingredients. Notable among these events is the introduction in 1985 of a fungal MBP product, 'mycoprotein', sold commercially for human consumption in England.

Microbial biomass proteins are potentially useful in supplementing the need for protein in animal and human nutrition. In addition, the production of MBP from waste residues and surplus raw materials could provide economic control of some forms of environmental pollution resulting from various industrial and agricultural operations and, concurrently, alleviate some of the global malnutrition and hunger problems. Governments and commercial enterprises are interested in all aspects of these potentials which could have far-reaching socio-economic benefits worldwide.

How safe are MBP products? What are their nutritional values? Are they economical to produce? How are government regulatory bodies involved in their commercialization? What are the market opportunities? These and other questions about MBP products are addressed in this book by some of the world's foremost experts in the field, including contributions from representatives of both developed and developing countries. The book is aimed at students, researchers and policy-makers in industry, government and academia who are interested in the resolution of problems in MBP commercialization.

The material, which ranges from basic scientific principles to practical engineering design and economic considerations, is treated in two sections:

Process Development and Economic Factors, and Product Safety and Nutritional Factors. Both these factors are crucial to the eventual commercial success of an MBP process as exemplified by the story of mycoprotein discussed in this work. This story marks a milestone in the history of the food industry which has been traditionally a very conservative one. In addition, this story draws attention to the importance of process engineering criteria; especially noteworthy is that a so-called toxigenic organism is 'controlled' in a continuous process to produce a safe, non-toxic, food-grade product. It is interesting to note that there are over one hundred MBP plants currently in operation in Eastern European countries, especially the USSR, and only a relatively few elsewhere.

In the preparation of this volume we are grateful for the professional assistance from Terri-Lee Schmidt for typing and Chris Krebs for proof-reading of the manuscripts. The National Research Council of Canada and UNESCO provided support for the publication arrangements.

MURRAY MOO-YOUNG
KENNETH F. GREGORY

CONTENTS

LIST OF CONTRIBUTORS

F. ARTEMI
Istituto di Chimica Agraria, Universitá degli Studi di Viterbo, 01100 Viterbo, Italy

M. BADIANI
Istituto di Chimica Agraria, Universitá degli Studi di Viterbo, 01100 Viterbo, Italy

P. C. BELL
School of Administration, University of Western Ontario, London, Ontario N6A 3K7, Canada

H. G. BOTTING
Health and Welfare Canada, Bureau of Nutritional Sciences, Food Directorate, Health Protection Branch, Tunney's Pasture, Ottawa, Ontario K1A 0L2, Canada

J. G. BUCHANAN-SMITH
Department of Animal and Poultry Science, University of Guelph, Guelph, Ontario N1G 2W1, Canada

J. D. BU'LOCK
Weizmann Microbial Chemistry Laboratory, University of Manchester, Oxford Road, Manchester M13 9PL, UK

E. R. CHAVEZ
Department of Animal Science, Macdonald College, McGill University, Ste Anne de Bellevue, Québec H9X 1C0, Canada

O. R. CONTRERAS
Microbiology Laboratory, Industrial Microbiology Department, CENIC, PO Box 6990, Havana, Cuba

M. FELICI
Istituto di Chimica Agraria, Universitá degli Studi di Viterbo, 01100 Viterbo, Italy

L. B. FLORES COTERA
Department of Biotechnology and Bioengineering, DINVESTAV-IPN, PO Box B 14-740, Mexico DF, Mexico

G. GIOVANNOZZI SERMANNI
Istituto di Chimica Agraria, Universitá degli Studi di Viterbo, 01100 Viterbo, Italy

R. GUAY
Groupe de Recherche en Recyclage Biologique et Aquaculture, Centre de Recherche en Nutrition, Université Laval, Ste Foy, Québec G1K 7P4, Canada

D. O. HITZMAN
Phillips Petroleum Company, Bartlesville, Oklahoma 74005, USA

S. K. HO
Feed and Fertilizer Division, Agriculture Canada, Ottawa, Ontario K1A 0C6, Canada

C. R. JONES
Department of Animal and Poultry Science, University of Guelph, Guelph, Ontario N1G 2W1, Canada

N. KOSARIC
Department of Chemical Engineering, University of Western Ontario, London, Ontario N6A 3K7, Canada

J. H. LITCHFIELD
Battelle Columbus Laboratories, 505 King Avenue, Columbus, Ohio 43201, USA

M. Luna
Istituto di Chimica Agraria, Universitá degli Studi di Viterbo, 01100 Viterbo, Italy

G. K. Macleod
Department of Animal and Poultry Science, University of Guelph, Guelph, Ontario N1G 2W1, Canada

M. Moo-Young
Department of Chemical Engineering, University of Waterloo, Waterloo, Ontario N2L 3G1, Canada

D. N. Mowat
Department of Animal and Poultry Science, University of Guelph, Guelph, Ontario N1G 2W1, Canada

S. Natori
Meiji College of Pharmacy, Tokyo, Japan

J. F. Nobile
ITT Rayonier, Stamford, Connecticut, USA

J. de la Noue
Groupe de Recherche en Recyclage Biologique et Aquaculture, Centre de Recherche en Nutrition, Université Laval, Ste Foy, Québec G1K 7P4, Canada

R. W. Peace
Health and Welfare Canada, Bureau of Nutritional Sciences, Food Directorate, Health Protection Branch, Tunney's Pasture, Ottawa, Ontario K1A 0L2, Canada

Y. Pouliot
BIONOV CNP Inc., Québec, Canada

D. Proulx
Groupe de Recherche en Recyclage Biologique et Aquaculture, Centre de Recherche en Nutrition, Université Laval, Ste Foy, Québec G1K 7P4, Canada

M. Raices
Microbiology Laboratory, Industrial Microbiology Department, CENIC, PO Box 6990, Havana, Cuba

G. Sarwar
Health and Welfare Canada, Bureau of Nutritional Sciences, Food Directorate, Health Protection Branch, Tunney's Pasture, Ottawa, Ontario K1A 0L2, Canada

N. S. Scrimshaw
Clinical Research Center, Massachusetts Institute of Technology, 50 Ames Street, Cambridge, Massachusetts 02114, USA

S. Sekita
National Institute of Hygienic Sciences, 1-18-1 Kamiyoga, Setagaya-ku, Tokyo 158, Japan.

G. L. Solomons
RHM Research Centre, Lincoln Road, High Wycombe, Bucks HP12 3QR, UK

K. H. Steinkraus
Institute of Food Science, New York State Agricultural Experiment Station, Cornell University, Geneva, New York 14456, USA

M. de la Torre-Louis
Department of Biotechnology and Bioengineering, DINESTAV-IPN, PO Box B 14-740, Mexico DF, Mexio

S. P. Touchburn
Department of Animal Science, Macdonald College, McGill University, Ste Anne de Bellevue, Québec H9X 1C0, Canada

J. Turcotte
Département de Chimie, Université Laval, Ste Foy, Québec G1K 7P4, Canada

S. Udagawa
National Institute of Hygienic Sciences, 1-18-1 Kamiyoga, Setagaya-ku, Tokyo 158, Japan

J. N. Udall
Clinical Research Center, Massachusetts Institute of Technology, 50 Ames Street, Cambridge, Massachusetts 02114, USA

1

TECHNICAL ECONOMIC AND MARKET STRATEGIES FOR MICROBIAL BIOMASS PROTEINS

John H. Litchfield
Battelle Columbus Laboratories
Columbus, Ohio
U.S.A. 43201

INTRODUCTION

What are the important technical, economic and market considerations in developing microbial biomass protein (MBP) products? To answer this question, we will discuss applications of MBP process factors including raw materials, product utility for food or feed applications, markets, economic and regulatory considerations, which are common to MBP products, from both photosynthetic and nonphotosynthetic micro-organisms although emphasis will be placed on nonphotosynthetic processes. Details on specific processes are covered in recent reviews (Batt and Sinskey, 1984; Litchfield, 1983a, b, 1984; Tanaka and Matsuno, 1985) and in other papers presented at this meeting, (Ban and Glanser-Soljan, 1985; Graille et al., 1985; Guiraud and Galzy, 1985; Moo-Young, et al., 1985, Nobile, 1985).

APPLICATIONS OF MBP

Table 1 summarizes MBP product values starting from raw materials and ranging from the primary product, microbial cells, to a variety of added value products. For the purpose of this discussion, we will consider food or feed uses of MBP. In the case of food, the products can be based directly on MBP or can be processed further to improve acceptability. The term "acceptability" includes sensory, nutritional, functional and safety aspects of the product in either human food or animal feed applications. Further processing includes texturization by addition of functional food additives, spinning into fibers, or extrusion, blending with flavorants, making protein concentrates and isolates by disrupting cells, removing cell walls and nucleic acids, or by preparing autolysates or hydrolysates to yield peptides and amino acids. MBP products designed for feed applications can be used to replace protein ingredients such as oil seeds or fish meal or as an additive to other plant or animal protein ingredients. We shall consider the performance of MBP in animal feeds subsequently in this paper.

TABLE 1: Microbial biomass product values

Raw Materials	Primary Products	Added Value Products
Carbohydrate (Sugars, Starch, Cellulose Hemicellulose)	Dried Microbial Cells	Microbial Protein Concentrates
		Microbial Protein Isolates
Alcohols (Methanol, Ethanol)		Nucleic Acids
		Amino Acids
		Pigments
		Vitamins
		Polysaccharides
		Lipids and Steroids
		Enzymes

It is important to make a decision on the desired product application at the outset of the development program. In the United States, facilities for manufacturing food-grade MBP products must operate under the Food and Drug Administration's Good Manufacturing Practices regulations and the products must meet FDA requirements for safety. Similar conditions apply in most countries. Feed grade MBP products can be manufactured under less stringent conditions than food products, but must meet regulatory agency requirements for safety including freedom from microbial or plant toxins, heavy metals and toxic chemical residues. (Food and Drug Administration, 1984).

PROCESS CONSIDERATIONS

The major steps in typical MBP processes based on nonphotosynthetic microorganisms are: raw materials, treatment, bioreactor, product separation, and product purification. I shall discuss these steps from the standpoint of their impact on process economics.

Raw Materials

Raw materials requirements for MBP production are governed by the requirements for growth and product formation which usually include: carbon and an energy source, a nitrogen source, O_2, minerals and supplementary nutrients. At the Symposium on Biomass Conversion Technology held at the University of Waterloo in 1984, I presented some of the considerations in selecting raw materials for MBP production including availability, composition and physical characteristics, performance and costs(Litchfield, 1984).

Here, I shall emphasize the raw materials for MBP processes based on nonphotosynthetic micro-organisms. Table 2 shows materials requirements for selected classes of MBP processes based on bacteria, yeasts and fungi. Production media should be developed on the basis of cell composition. Haggstrom (1985) has shown that the elemental composition of typical growth media for bacterial cells reported in the literature often deviates widely from that of the cells themselves (Table 3).

TABLE 2: **Materials requirements of microbial biomass protein processes**

Material	Quantity/Metric ton of MBP			
(Metric tons)	n-Paraffins Yeast	Methanol Bacteria	Ethanol Yeast	Carbohydrate Yeast, Fungi
Carbon and Energy Source	0.87-1.05	2.0	1.4	2.00
Ammonia	0.14	0.13-0.16	0.09	0.09
Phosphoric acid (100% Basis)	0.05-0.08	0.095	0.05	0.06
Mineral Nutrients (Fe, K, Mg, Mn, Zn)	0.02	0.03	0.03	0.03

The values for carbohydrates given in Table 2 are based on the assumption that the carbohydrate supplied is in a form that is assimilated by the growing cells. Tanaka and Matsuno (1985) discuss pretreatment of lignocellulosic materials to make them suitable for MBP production. It is clear that only a portion of such substrates can be converted to utilizable form and amounts of these raw materials required per unit weight of MBP are considerably greater than that shown in Table 2.

Table 4 shows that prices for selected carbohydrate substrates for MBP production decreased markedly over the 1980-1985 period. As shown in Table 5, the price of anhydrous ammonia also decreased, but the price of 85 percent phosphoric acid increased over this same period. Current 1985 prices for ethanol and methanol (100 percent basis) are approximately $0.57/kg and 0.24/kg, respectively.

TABLE 3: **Composition of bacterial cells and growth media**

Elements	Cells	Media
N	100	100
P	23	176
K	14	201
S	8.9	59
Mg	4.9	15
Na	3.2	66
Ca	3.0	11
Cl	2.5	123
Fe	0.3	2.2
Zn	0.14	0.13
Cu	0.03	0.04
Mn	0.05	0.15
Co	0.003	0.02
Mo	0.002	0.09
B	0.006	0.01

Haggstrom, 1975

TABLE 4: **Price trends for selected raw materials for microbial biomass production.**

Raw Material	Price U.S. Dollars/kg 1980	1985
Glucose (Dextrose) Hydrate	0.64	0.53
Sucrose (Cane), raw		
U.S.	0.68	0.46
World	0.09	0.06
Molasses, Cane	0.18	0.07

TABLE 5: **Price trends for supplemental nutrients for microbial biomass production.**

Nutrients	Price U.S. Dollars/kg 1980	1985
Ammonia, Anhydrous	0.17	0.15
Phosphoric Acid 85%	0.52	0.74

The costs of wastes that are suitable as substrates for MBP production are considerably lower than those of commercially available raw materials. However, costs for collecting, transporting and pretreating waste substrates must be considered in determining whether or not it is economically feasible to use these materials for MBP production at a given site. In general, carbon and energy source costs may range from 14 percent of total manufacturing costs for wastes to greater than 50 percent for highly purified substrates (Litchfield, 1983a).

We must also consider future changes in the availability of wastes as substrates for MBP production. Large quantities of cheese whey, a source of lactose, have been available in major cheese producing regions. Recently however, the feasibility of on-farm ultrafiltration of milk to yield a precheese suitable for manufacturing has been demonstrated at an 800 head dairy farm in California (Zall, 1985). The permeate was kept on the farm to supplement animal feeds. Expansion of this on-farm production could reduce the availability of lactose-containing substrates such as cheese whey or permeate in the future.

MBP Production and Product Recovery Process

The use of various bioreactor designs, such as the conventional baffled, agitated tank and airlift fermenters in MBP processes has been covered previously (Litchfield, 1983a,b, 1984). Recent work at Nagoya University in Japan on automated control of mineral ion concentrations in fed batch culture of *Candida brassicae* with ethanol and ammonia feeds achieved cell mass concentrations of 138 g/L (dry weight basis), 150 g/L with *Candida utilis* and 140 g/L with *Kluyveromyces fragilis* on concentrated whey permeate in a continuous process. Productivities are approximately 30 g/L•hr with *C. utilis*.

These high cell densities enables the fermentor broth to be spray-dried directly without centrifugation and eliminates waste streams. These developments could lead to significant reductions in capital and operating costs for MBP processes that can operate at these high cell densities and productivities.

We mentioned previously the possibility of preparing specialty MBP products, such as protein concentrates and isolates or texturized or formed MBP products. McNairney (1984) described an aqueous extraction process for co-production of a protein concentrate and nucleic acids from ICI's Pruteen product (*Methylophilus methotrophus*) grown on methanol. The nucleic acid content of the cell mass was reduced from 14 percent to less than 3 percent which meets the Protein Advisory Group (1975) recommended levels for microbial protein products suitable for human consumption.

The protein concentrate can be enzymatically hydrolyzed to yield a variety of functional protein products having desired foaming, solubility or emulsifying characteristics. Also, purified RNA can be recovered from the nucleic acid fraction for use in preparing flavoring nucleotides and products for cosmetic or pharmaceutical uses.

Hoechst-Uhde in West Germany, has prepared a 90 percent protein isolate called Probion from *Methylomonas clara*. Again, the nucleic acids removed during this process can be purified further as in the ICI process.

None of these protein concentrate products has been approved yet by regulatory agencies in the countries in which they have been developed. However, with the demonstration of safety to the satisfaction of regulatory agencies, one can envision the marketing of a variety of new functional protein ingredients.

PRODUCT UTILITY

Product utility includes both acceptability of MBP to users and nutritional value for human food or animal feeding applications. Some of the more important considerations in acceptability of MBP include stability, absence of (potential) toxic factors, processability, sensory factors (taste, odor), factors affecting absorption and digestion. These acceptability factors apply to both human food and animal feed products.

Nutritional value for humans or for livestock is determined best by actual performance in human clinical studies or in livestock feeding trials. This subject is discussed in detail in other papers in this monograph.

ECONOMIC-MARKET CONSIDERATIONS

Tables 6, 7 and 8 present comparisons of 1985 selling prices for MBP and plant and animal protein products, respectively. It is clear that MBP products sell for significantly higher prices than other plant or animal protein products. These higher prices of MBP products are based on either additional nutritional values due to B-vitamin contents or on functional effectiveness as an ingredient in food products.

TABLE 6: Comparison of selling prices for microbial protein products

Product, Substrate, and Quality	Crude Protein Content N x 6.25 (Dry Wt., Basis)	1985 Price Range U.S. $/kg (Dry Wt. Product Basis)
Candida utilis		
Food Grade	52	2.64-2.75
Kluyveromyces fragilis		
Cheese Whey, Food Grade	54	2.40-2.51
Saccharomyces cerevisiae		
Brewers, Debittered, Food Grade	52	2.42-2.45
Feed Grade	52	0.42-0.62

TABLE 7: Comparison of selling prices for plant protein products

Product	Crude Protein Content N x 6.25 (Dry Wt., Basis)	1985 Price Range U.S. $/kg (Dry Wt. Product Basis)
Alfalfa, Dehydrated	17	0.08-0.09
Soybean Meal, Defatted	49	0.15-0.16
Soy Protein Concentrate	70-72	0.95-0.99
Soy Protein Isolate	90-92	2.64-2.77

TABLE 8: Comparison of selling prices for animal protein products

Product	Crude Protein Content N x 6.25 (Dry Wt., Basis)	1985 Price Range U.S. $/kg (Dry Wt. Product Basis)
Fish Meal, Menhaden	60	0.33-0.34
Meat and Bone Meal	50	0.16-0.17
Dry Skim Milk	37	1.03-1.06

It is interesting to note the world trends in soybean production given in Table 9. Production has decreased in the USA in the period 1981-1983 resulting from adverse growing conditions and a reduction of acreage. Brazilian production has fluctuated but will undoubtedly increase in the future. If the Peoples Republic of China increases soybean production in the future, beyond domestic needs, additional soybeans could become available in the world market.

TABLE 9: World trends in soybean production

Region	Production (1000 metric tons)			
	1974-76	1981	1982	1983
World	58,076	88,155	92,058	77,541
USA	36,711	54,136	59,611	42,639
Brazil	9,666	15,007	12,835	14,852

FAO (1984)

The recent decrease in world soybean production has led to record high prices for soybean oil in the range of $0.66/kg but depressed prices for soybean meal. This depressed price of soybean meal makes it difficult for MBP products to compete in animal feedstuff markets, as ICI has learned from attempts to market its Pruteen product as an animal feed in Western Europe.

Table 10 presents the key economic-market considerations in developing MBP products for either human food or animal feed applications. These questions should be raised at the outset of any MBP development program.

TABLE 10: Economic market considerations

- Who are the potential buyers
- Where are buyers located
- What do buyers want - type of product, quality, quantity
- What price will buyers pay
- Who are competitors - companies, location and products
- Will extensive market development, and sales service be needed

The answers to these questions can then be compared with the answers to the following:

1. Which applications have the greatest potential food or feed? Market size, potential annual volume of production?
2. What substrates are available? Costs?
3. What are the most appropriate MBP technologies, micro-organisms, processes,

scale, potential capital and operating costs?

4. What is the regulatory status of the proposed micro-organisms and products? Will extensive safety studies be required?
5. What is state of development of the most appropriate MBP technologies-laboratory, pilot plant or commercial?
6. What are the potential costs for research and for pilot plant and commercial development?
7. Will licenses be needed on patented MBP technologies and products before proceeding with commercial development?

This approach should provide a useful framework for MBP product and process decision-making. This strategy of defining potential market applications and then comparing available MBP technologies meeting the potential markets should avoid investments in processes that are unlikely to be economically viable.

REFERENCES

1) Ban, S.N. and Glanser-Soljan, M. 1985. Two-step processing of waste food industry and agriculture with simultaneous production of energy and valuable food products. Private communication.

2) Batt, C.A. and Sinskey, A.J. 1985. Use of biotechnology in the production of single-cell protein. Food Technol. 38(2), 108-111.

3) Food and Drug Administration. 1984. Guideline for New Animal Drugs and Food Additives Derived from a Fermentation: Human Food Safety Evaluation, U.S. Department of Health and Human Services, Washington, DC.

4) Graille, J., Ratomahenina, R., Pina, M. and Galzy, P. 1985. Production of food yeast from lipids. Private communication.

5) Guiraud, J.P. and Galzy, P. 1985. Protein production from Jerusalem Artichoke. Private communication.

6) Haggstrom, L. 1975. Ph.D. Thesis, Unviersity of Lund, Sweden.

7) Litchfield, J.H. 1983a. Single cell proteins. Science 219: 740-746. Litchfield, J.H. 1983b. Technical and economic prospects for industrial proteins in the coming decades, pp. 9-27 in International Symposium on Single Cell Proteins Technique et Documentation, Paris, France.

8) Litchfield, J.H. 1984. Production of foods, food additives and feeds from biomass by microbiological processes. Paper 9-1, Symposium on Biommass Conversion Technology, University of Waterloo, Ontario, Canada, July 16-20.

9) McNairney, J. 1984. Modification of a novel protein product. J. Chem. Technol. Biotechnol. 34B: 206-214.

10) Moo-Young, M., Burrell, R. and Scharer, J.M. 1985. Waterloo fungal and yeast SCP process developments. Private communication.

11) Nobile, J. 1986. The "Raypro" process. Private communication.

12) Shay, l.K. and Wegner, G.H. 1984. Production of torula yeast by improved fermentation technology. Paper 110, 44th Annual Meeting Institute of Food Technologists, Anahein, California, June 10-13.

13) Shay, L.K. and Wegner, G.H. 1985. Non-polluting conversion of cheese whey permeate to food yeast. Paper 27, 45th Annual Meeteing, Institute of Food Technologists, Atlanta, Georgia, June 9-12.

14) Suzuki, T., Mori, H., Yamane, T. and Shimizu, S. 1985. Automatic supplementation of minerals in fed-batch culture to high cell mass concentration. Biotechnol. Bioeng. 27:192-201.

15) Tanaka, M. and Matsuno, R. 1985. Conversion of lignocellulosic materials to single-cell protein (SCP): recent developments and problems. Enzyme Microb. Technol. 7:197-206.

16) Zall, R.R. 1985. On-farm ultrafiltration of milk: a national study. Paper 324, 45th Annual Meeting, Institute of Food Technologists, Atlanta, Georgia, June 9-12.

11

RAYPRO: A CASE HISTORY OF SINGLE CELL PROTEIN DEVELOPMENT

J.F. Nobile
ITT Rayonier
Stamford, CT
U.S.A.

INTRODUCTION

Usually there are two sources for new product ideas. The first is a company's R&D department who, working independently, develops a technology for a new product. In this case the concept is presented to marketing with the challenge..."We can make this! Can you sell it?" The second source is the company's marketing group who reading certain signals from the field determine the market need for a new product. In this case, R&D is presented with the challenge..."We can sell this! Can you make it?"

The development project I will be discussing here received its conception from a third and totally different source. That source was the US Environmental Protection Agency (EPA) who essentially presented ITT Rayonier with the challenge..."You must make it! And you must sell it". This challenge was presented because of changes in effluent disposal regulations that reduced the volume of waste material that could legally be discharged by one of our mills in the state of Washington. The solution to this problem allowed the mill to continue operations while remaining in compliance with these new regulations and resulted in the development of a new and profitable chemical product for the company.

THE PROCESS RATIONALE

Let me state the problem in more detail. The ITT Rayonier mill in Port Angeles, Washington, uses an ammonia-based acid sulfite pulping process to produce several grades of dissolving and paper pulps. Purification of the cellulose pulp fibers occurs in a series of pulping and bleaching operations and comprise the mill effluent to be treated. The mill's National Pollutant Discharge Elimination System Permit was revised by the EPA and established reduced limits on the amount of organic material in the mill wastewater that could be discharged into the straits of Juan de Fuca of Puget Sound. To meet these permit requirements, it was necessary for Rayonier to construct a biological secondary treatment system to achieve over 90% biological oxygen demand removal.

In this system, micro-organisms are employed to consume organic materials in the mills wastewater. Approximately one-half million gallons of excess micro-organisms are produced each day in this process and must be removed from the system. This wasted biomass can be an odiferous and unmanageable material, especially considering that it is composed primarily of water. It became necessary for us to identify an environmentally satisfactory method for disposal of this material.

Our Operations Group, R&D Staff, Engineering Division, Marketing Group, and in independent consultant evaluated a number of possible secondary waste disposal schemes including: landfilling, ocean disposal, incineration, processing for sale. A number of considerations led us to conclude early in the project that a new disposal approach was desirable. Conventional approaches such as landfilling, ocean disposal, and incineration offered drawbacks which in our case would be magnified due to the large volume of waste to be disposed of (enough to cover one acre of land to a depth of four inches each day). Landfilling this quantity of waste would make large amounts of land unavailable for productive use, would require special protection against groundwater and surface contamination and could cause odor and other aesthetic problems.

Waste disposal by incineration would be costly because of the high moisture content of the material and Ocean Disposal would also be costly and create logistical problems. Therefore, a high priority was assigned to the goal of converting this waste material into a useful and profitable product for the company.

A study performed by R&D indicated that this waste stream could be converted into an animal feeds protein supplement and a manufacturing process was identified and evaluated on a laboratory-scale. Marketing confirmed the fact that a suitably sized market existed for protein supplements, and the proposal was presented to management. Once the management decision was made to convert these single-cell organisms to an animal feed protein supplement, two other major tasks remained.

In the first case, we had to adapt and operate a technology never before used in the pulp and paper industry to manufacture a product that met extremely strict government standards for animal feed. As illustrated in Figure 1, the ITT Rayonier SCP production system begins with micro-organisms from the secondary treatment plant which float to the surface of treatment tanks and are skimmed from these tanks and pumped to the drying plant. In the drying plants, these organisms are first mixed with polymers to allow water to drain more easily from their surfaces. This conditioned sludge is then spread among five "Tait Andritz" absorbent belt filters. At this point the moisture content is reduced from 98 to 88%. A metered quantity of tallow is then added to the product cake and the mixture is passed through a series of multiple-effect, falling film evaporators. This evaporation process is known as the Carver-Greenfield process which is a patented technique for the removal of water by evaporation without the problems of thickening, scaling, and fouling that are often associated with less sophisticated processes. This process license is owned by "DeHydro Tech" of East Hanover, New Jersey.

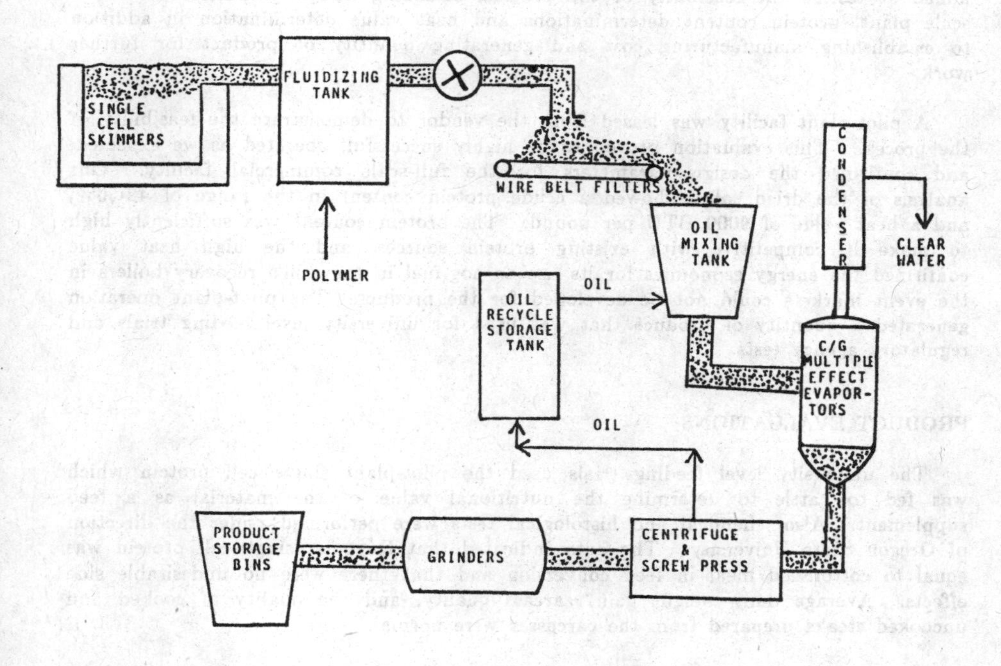

FIGURE 1: ITT Rayonier SCP manufacturing process

After the water is evaporated, the solids remain suspended in the tallow as a fluid slurry. The tallow is removed by centrifugation and press filtration, and returned to the feedtank for recycling. The dried product is ground to reduce it to a powder and conveyed to storage bins for shipment. The resultant product is discharged as 83% dry material, 16% tallow, and 1% moisture.

The second major task we faced in order to have a successful venture depended heavily on an intensive effort by marketing to gain market acceptance of a new and unfamiliar source of high quality protein.

Several pre-marketing programs were undertaken to qualify the Port Angeles single-cell protein as an animal feed supplement. In addition, market studies were undertaken to confirm financial and market justification for the project.

These pre-marketing qualification programs included pilot scale evaluations, university level feeding trials, and regulatory agency testing. These evaluations were aimed at testing the feasibility of the process, obtaining design parameters of full - scale plant, protein content determinations and heat value determination in addition to establishing manufacturing cost and generating quantity of product for further work.

A pilot-plant facility was leased from the vendor to demonstrate the feasibility of the process. This evaluation proved to be highly successful, operated as we expected, and confirmed the design parameters for the full-scale commercial facility. Our analysis of the dried solids showed a crude protein content in the range of 48-50%, and a heat value of 9000 BTU per pound. The protein content was sufficiently high to make it competitive with existing protein sources, and the high heat value confirmed the energy economics for its use as hog-fuel in the mill's recovery boilers in the event markets could not be developed for the product. The pilot-plant operation generated a quantity of product that was used for university level feeding trials and regulatory agency tests.

PRODUCT EVALUATIONS

The university level feeding trials used the pilot-plant single cell protein which was fed to cattle to determine the nutritional value of the material as a feed supplement. Also, chemical and histological tests were performed under the direction of Oregon State University. The data indicated that Rayonier single-cell protein was equal to cottonseed meal in feed conversion and that there were no undesirable side effects. Average daily weight gain, carcass quality, and the quality of cooked and uncooked steaks prepared from the carcasses were normal.

Regulatory testing agencies used the pilot-plant material which was analyzed by laboratories of the Washington State Department of Agriculture. It was found to meet the requirements of that agency with respect to nutrient value for protein, fat and fiber, and absence of pathogenic bacteria, pesticides, PCB's and low levels of heavy metals. Sale of the product in Washington State required only their approval and registration with the American Association of Feed Control Officials. The Food & Drug Administration stated it would permit the sale of the product within the State of Washington so long as its use was under surveillance of that state's Department of Agriculture.

So here we were with a new product. The economics looked good, the manufacturing process was a viable one, the product displayed better than average nutritional qualities, and we had obtained the necessary approvals from the regulatory agencies to begin selling the product. Now the real challenge began...to gain market acceptance.

Our Marketing Research indicated that there was a ready market for animal feed protein supplements in the State of Washington, since most competitive products were imported from out-of-state sources.

As indicated in Table 1, the total yearly feed consumption in Washington State was roughly 1.1 million tons in 1980, and approximately 115,000 tons of protein feed supplements was used in those feed formulations. These protein supplements were mainly cottonseed and soybean meals from the midwest and Canola meal from Canada.

TABLE 1: Annual feed consumption and livestock and poultry population in Washington state (1980)

Species	Feed (Thousand Tons)	Protein (Thousand Tons)	Head Count
Layers	69	7	5,250
Broilers	89	9	20,641
Dairy	581	58	200
Cattle	325	33	1,250
Sheep	25	3	80
Pig	52	5	205
Total	1,138	115	

Pricing and marketability of this material would be contingent upon product analysis. With crude protein content being the most important factor. As shown in Table 2, Rayonier single-cell protein ("RayPro") compared favorably in comparison with competitive products and was priced at 80% of the soybean meal price to facilitate our market entry.

A further analysis (see Table 1) of the numbers of various types of animals-on-feed in the state indicated that we should direct our marketing efforts to the layer, dairy, and broiler feed areas.

The conclusions that were drawn from our marketing research studies were as follows: (1) Assuming an average production of 550 tons per month of single-cell protein all but two feed segments could individually sell-out the annual production from the mill. (2) We would need to achieve a 6% penetration of the total market for protein supplements to sell our entire production (this was our goal). (3) The importance of the dairy, layer, and broiler feed segments became obvious since they were among the largest segments in terms of feed consumed, and these herds were mainly located in western Washington close to our production facility.

TABLE 2: Comparative protein levels of feed products

	Content
RayPro	45-50%
Soybean Meal	48%
Cottonseed Meal	41%
Canola Meal	28%
Linseed Meal	33%
Fish Meal	60%

MARKETING

A market development plan was established to address two specific objectives: (1) To gain market acceptance for single-cell protein, and (2) To correct several previously identified product quality problems.

The first strategy used to meet the market acceptance objective was to perform feeding trials on commercial scale production of the product. All of our previous feeding trials had been performed using pilot-scale material and feedback from the field indicated that the majority of this work had to be repeated using material from the commercial facility. These new feeding trials were performed using the targeted animal types of dairy cows, layers, and broilers. A full-scale dairy feeding trial was conducted at Oregon State University for a 120 day period using a herd of 48 experimental and control cows.

Factors evaluated during the trial included: feed intake per day, milk production vs. feed intake, body weight change, milk fat content, milk protein content, milk quality-judged by a taste panel, and heavy metal content of the milk. Conclusions of this study were that our single-cell protein could replace typical protein supplements in rations fed to lactating dairy cows and showed advantages over typical protein supplements in terms of milk fat content and feed conversion (gallons of milk produced per pound of feed consumed).

Broiler and layer trials were performed at Washington State University. Layer trials were performed for a period of 143 days and evaluated such factors as: egg production (number of eggs laid/hen), feed consumed/dozen eggs produced, and egg weights. Conclusions to this study indicated that ITT Rayonier SCP could replace typical protein supplements in layer feed with no detrimental effects and that it had the advantage of reducing the total feed consumed per dozen eggs laid. Broiler trials indicated a deficiency in the available lysine content of the SCP and since this was a critical amino acid for broiler growth, these trials were stopped at midstream until adjustments could be made to our manufacturing process, and the available lysine content could be improved.

Additional chemical and nutritional analyses were performed on the material during these trials to generate an expanded and revised profile which would allow the product to become a part of a "least-cost" computer program used by nutritionists and feed mills to select feed ingredients for a total feed ration. In this program a nutritionist inputs the desired nutritional characteristics of a feed to be formulated, along with the prices and chemical and nutritional profiles of all available nutrients. The computer then uses this data to select the ratios of available nutrients that result in a feed ration meeting these nutritional characteristics and at the least cost. These chemical and nutritional profiles were essential for us to gain market acceptance since without them we could not be a part of this "least-cost" program.

Our direct selling efforts were aimed at three separate levels within the industry: the nutritionist, the feed mill operator, and the farmer. We discovered that each played an important role in the acceptance of this new protein source and that each had to be sold individually on its merits and quality.

Other strategies used to achieve our market acceptance objective concerned product literature, advertising, and promotion. The trade name "RayPro" was selected to establish brand recognition for the product (Ray from Rayonier, and Pro from protein).

Specification sheets (see Table 3), process description sheets, and technical data sheets summarized the results of our feeding trials and were prepared and distributed to farmers, feed mills and nutritionists as part of a direct mail campaign.

Finally, a technical symposium was held in Seattle where farmers, feed mill operators and nutritionists were able to listen to papers presented by representatives of Washington State University and Oregon State University describing the feeding trials that we had performed. Also presented were papers by our R&D Group discussing the technical aspects of our SCP, and by our mill engineers addressing the manufacturing process.

Simultaneous with our efforts to gain market acceptance, strategies were put into effect to correct several quality problems that had been identified. A task force was established of representatives from R&D, engineering, mill operations, quality control and marketing to address such issues as increasing available lysine content, eliminating clumping, reducing fat content, adding anti-oxidant, and pelletizing.

In the period of time between March 1981 and February 1982, we had put our development plan objectives into effect, had solved our quality problems, and were beginning to make a significant volume of sales.

By the end of 1982, the mill was producing approximately 8000 tons of "RayPro" per year, and we were selling all of this production. This development project not only put what would otherwise have been a waste stream into productive use but also avoided entirely the historical drawbacks of conventional disposal techniques. Compared to other alternatives, "RayPro" sales have eliminated high operating costs for disposal and has dramatically improved the viability of the Port Angeles mill.

TABLE 3: RayPro™ Specification Sheet for Animal Feed Additives, Single Cell Protein from Pulp Processing

RayPro is a single-cell protein (SCP) from cells of micro-organisms. It is produced from a process stream of a wood pulp mill. It is high in protein and fat making it an ideal source of protein and energy for single stomach and ruminant animals. In numerous feeding trials it has proven to be an ideal protein and energy substitute for soybean, cottonseed and rapeseed meals.

Typical Proximate Analysis
(guaranteed analysis in parenthesis)

Crude Protein	53%	(45% min.)
Fat	16%	(13% min.)
Fiber	11%	(16% max.)
Moisture	1%	(3% max.)
Ash	6%	

Typical Amino Acid Analysis

Amino Acid	% of Total Crude Protein
Alanine	9.7
Arginine	7.2
Aspartic Acid	10.4
Cystine	0.7
Glutamic Acid	13.4
Glycine	6.1
Histidine	1.8
Isoleucine	4.5
Leucine	7.7
Lysine	5.4
Methionine	2.0
Phenylalanine	3.6
Proline	4.8
Serine	4.8
Threonine	5.7
Tryptophane	2.2
Tyrosine	3.2
Valine	6.8

CONCLUDING REMARKS

In 1982, the "RayPro" process received the American paper Institute/National Forest Product Association's "Environmental Improvement & Energy Management Award" in the solid waste management category. To date, we are continuing to successfully sell the entire production from the mill, and "RayPro" has become a valuable animal feed protein source in the State of Washington.

19

MICROBIAL PROTEINS AND REGULATORY CLEARANCE FOR RHM MYCO-PROTEIN

G.L. Solomons
RHM Research Ltd.
Lincoln Road
High Wycombe
Bucks. HP12 3QR
UK

Any established, ethical food company about to consider the production and sale to the public of a totally novel food produced by a totally novel process will be concerned above all else that the product is safe for human consumption. Even if no regulatory requirements exist, companies have moral obligations, which they take very seriously and are also rightly concerned that any hint of nutritional inadequacy or toxicological or immunological problem, could or would adversely affect the sales of their existing products or in the extreme case expose them to action for damages in the Courts of Law. No company is likely to view such potential threats lightly.

Therefore, the safety evaluation of a novel food must be the single most important part of the whole R & D programme (Edelman, Fewell & Solomons, 1983). A number of countries have, over the course of the last twenty to thirty years, elaborated and refined a comprehensive test procedure for examining the toxicological status of a range of chemical compounds, usually drugs, agricultural chemicals (pesticides, herbicides etc.), other environmentally hazardous chemicals and food additives (preservatives, anti-oxidants, colors, etc. (FDA, 1982)). Almost all of these compounds have known chemical composition, and most possess clearly definable chemical and physical properties; moreover, they are expected to be used, especially the food additives, at quite low concentration. These characteristics immediately distinguish these materials from a foodstuff, which is invariably of a complex structure, containing many different types of compounds (e.g. carbohydrates, proteins, lipids etc.), and intended for consumption at levels that could reach perhaps 100g/day. Since the main rationale for testing chemicals for safety relies on examining their effects at 100 times their maximum intended dose, it is immediately apparent that a food cannot be tested at such a level. The Protein Advisory Group of FAO/WHO has recently issued updated guidelines on the production and testing of Single Cell

Protein for human consumption (PAG/UNU, 1983 a,b,). The main differences between drugs and food with regard to their toxicological testing are shown in Table 1.

TABLE 1: Main differences between drug and food toxicity testing

	Drugs	Food
1	Methods well defined	Methods poorly defined
2	Test material usually a simple, chemically precise substance	Test material complex mixture of many chemical compounds
3	Highest dose level likely to produce toxic effect	Must not see toxic effects at any level of administration
4	Small physical dose	High physical dose
5	Easy to give excessive dose	Excessive dose difficult
6	Acute effects obvious	Acute effects difficult to produce
7	Generally independent of nutrition	Nutrition dependent
8	Metabolism relatively simple to follow	Complex metabolism
9	Cause/effect relatively clear	Cause/effect may be confused

When we were faced with providing evidence of safety for our novel protein product, produced by the aseptic fermentation of food grade glucose by a filamentous organism, ***Fusarium graminearum*** with the generic name myco-Protein (MCP), we decided that we would have to incorporate the material into animal diets at the highest possible concentration, compatible with appropriate nutritional requirements and with normal physiological function. For that reason, we had tested and rejected the use of highly concentrated extracts, since uncertainty of extraction, risk of any residual solvents and unphysiological concentrations of components causing osmotic and other effects in test animals render the results meaningless. We therefore formulated animal diets containing MCP up to a concentration of 54% dry weight thereby providing all of the protein supply (Duthie, 1975).

In principle, we decided that the normal spectrum of toxicological tests and data used for evaluating chemicals would need to be examined. Such a range is shown in Table 2 and covers sub-chronic studies of 28 and 90 days in rats and 90 days in baboons; life-span study, with *in utero* exposure for chronic toxicity and carcinogenicity, a multi-generation study (4 generations), teratology in both rats and rabbits, skin and eye sensitivity tests in rabbits and guinea pigs. The analysis and/or data required from these studies are indicated in Table 3. In addition to these conventional laboratory studies, we sought to gain confirmation of safety by using less conventional trials. For example, a 60 week egg production trial using laying hens, which are known to be sensitive to nutritional inadequacy as well as the presence of some toxic agents. Other animal trials have included neo-nate studies with calves and oestrogen potential studies in pigs.

TABLE 2: Suggested programme for toxicological evaluation of foods

Catagory	Duration and other features	Test species
Sub-acute	1, 3, 6 months	Rats, dogs, primates
Chronic toxicity and carcinogenicity	2 years or life-span	Rats, mice
Reproduction		
Multigeneration	At least 3 generations	Rats, mice
Teratology	Pregnancy	Rats, rabbits
Dominant lethal	Reproduction cycle	Rats
Mutagenicity	Various cytological systems	Mice

TABLE 3: Analysis and/or data required from toxicological tests

Study	Analysis/data
Acute/chronic toxicity	a) Food and water intake
	b) body weight
	c) haematology
	d) clinical chemistry
	e) macroscopic examination
	f) histopathology
	g) organ weights
	h) opthalmoscopy
	i) statistics
Reproduction	
Multi-generation	a) Food and water intake
	b) body weight
	c) mating performance
	d) pregnancy rate
	e) gestation period
	f) reproduction life-span
	g) pup mortality during lactation
	h) pup weight at 21 days of age
	i) gross pathology of six week old offspring
Teratology	a) Food and water intake
	b) body weight
	c) viable young
	d) embryonic death
	e) *corpora lutea*
	f) implantations
	g) pre-implantation losses
	h) foetal losses
	i) litter weight
	j) mean pup weight
	k) abnormalities: major malformation, skeletal anomalies

None of these studies showed responses that could be considered toxicological in origin. We therefore proceeded to tests in humans, looking in particular for immunological responses. During the course of our testing, approximately 4500 persons consumed the material on more than one occasion. Only one proven case of intolerance was established, the subject was sick and felt unwell but gave a negative prick test response to extracts of MCP; such responses to foods can often be prostaglandin mediated (Buisseret, et al. 1978), rather than necessarily of an immunological nature. Studies on atopics have also been carried out again with no adverse effects.

In addition to these toxicological studies, animal nutrition trials have been conducted to measure NPU, PER, slope ratios etc. (Table 4). The results of some of these studies are shown in Table 5.

TABLE 4: Measurement of nutrition value

Study	Assay	Animal species
Protein quality	Digestibility	Rats
	NPU	Rats
	PER	Rats
	Slope ratio	Rats
	Available amino acids eg. lysine, methionine	Chicks, rats
Energy	Digestible	Chicks, rats
Availability	Metabolisable	Chicks, rats
Trace nutrients	Vitamins and minerals	Chicks, rats, etc.
Feeding trials	Various	Rats, chickens, pigs, calves, etc.

TABLE 5: Nutrition values for MCP

a) animal trials b) human studies

a)

Rat assays	NPU	PER
MCP	61	2.4
MCP & methionine	82	3.4
Casein	70	2.5

Chick assay	
Metabolisable energy	3.0 Kcal/g

b)

Human study	Biological Value
MCP	84
Milk protein	85

As well as animal trials, extensive chemical analyses on both the fermentation process and the products (both before and after the RNA reduction stage) have confirmed the nutritional adequacy of the material. Tables 6, 7 & 8 illustrate the proximate analysis, nitrogen composition and amino acid composition of MCP. Moreover, extensive testing for a range of mycotoxins have all proved negative, even where detection levels were as low as 1 μg/kg.

TABLE 6: Proximate analysis of MCP

	% of dry weight
Crude protein (N x 6.25)	54-59
True protein (α-amino N x 6.22)	47-50
RNA	1- 2
Total lipid	12-13
Crude fibre	16-21
NDF/Dietary fibre	22-28
Ash	3- 3.5
Water soluble sugars	2- 4

TABLE 7: Nitrogen distribution of a) before and b) after RNA reduction

	% Dry weight	
	a)	b)
Free amino-acid	0.7	-
Protein-N	6.3	7.7
Nucleotide N	0.2	-
RNA N	1.5	0.1
n-acetyl glucosamine N	1.0	1.5
total	9.7	9.3
TN (by analysis)	9.7	9.0

TABLE 8: Amino acid composition of MCP

Amino acid	g/100g amino acid
Lysine	8.1
Methionine	2.2
Cysteine	0.8
Threonine	5.5
Tryptophan	1.7
Valine	5.9
Leucine	8.3
iso-Leucine	5.1
Phenylaline	4.8
Histidine	3.9
Arginine	7.7
Tyrosine	3.9
Aspartic acid	9.9
Serine	5.1
Glutamic acid	11.5
Proline	4.7
Glycine	4.6
Alanine	6.3
	100.0

We have also examined the bacteriological safety of the material comparing it with chicken and fish when inoculated with a range of food poisoning organisms. These tests showed that MCP has a slightly less propensity to support the growth of such organisms.

During the course of our test programme, which has lasted in all for sixteen years, we have fed approximately 100 te dry weight of MCP. As a result of all of this testing, we have been given clearance by the UK regulatory authority, Ministry of Agriculture, Fisheries & Food, for the sale of products to the public. The first product appeared on the supermarket shelf 21 years after the first experiments were initiated.

REFERENCES

Buisseret, P.D., Heizlemann, D.I., Youlton, L.J.F. & Lessof, M.H. (1978). Prostaglandin synthesis inhibitors in prophylaxis of food intolerance. *The Lancet*, April 29th, 906-908.

Duthie, I.F. (1975). Animal feeding trials with a microfungal protein *in* 'Single-Cell Protein II', eds. Tannenbaum, S.R. & Wang, D.I.C., MIT Press, Cambridge, Mass. pp 505-544

Edelman, J., Fewell, A. & Solomons, G.L., (1983), Myco-Protein-a new food. *Nut. Abs. & Revs. in Clin. Nut.*, 53. No. 6., 471-480

Fawell, J. (1979). Cadmium/Phosphorus Nutrition-Nephrocalcinosis in Rats. Paper, Imp. Cancer Res. Fund, Lab. Animals Service Nutr. Group Meet. Nov 26th.

PAG/UNU (1983a) Guideline No.12: The production of single-cell protein for human consumption. *Fd. & Nut. Bull.* 5 No.1., 64-66.

PAG/UNU (1983b). Guideline No. 7: Human testing of novel foods, *Fd. & Nut. Bull.* 5, No.2., 77-80.

US Food & Drug Admin., Bureau of Foods (1984). Toxicological Principles for the Safety Assessment of Direct Food Additives & Additives used in Food. FDA, Washington, USA.

THE PROVESTEEN PROCESS - AN ULTRA-HIGH DENSITY FERMENTATION

D.O. Hitzman

Senior Research Associate, Biotechnology
Phillips Petroleum Company
Bartlesville, Oklahoma
U.S.A. 74005

ABSTRACT

The Provesteen ultra-high cell density fermentation process was developed at Phillips Petroleum Company in response to a need to produce large amounts of protein biomass. The process, which involves the continuous production of yeast at cell densities of 120-150 g/L dry weight, initially used methanol as feedstock. This technology has been expanded to include feedstocks such as sugars, molasses, whey, etc. and can employ other microbial species. These developments opened other new commercial applications for this novel SCP technology. Comparative figures for the Phillips-Provesta processes and product will be presented. This technology has been extended to demonstrate that practical fermentor production of rDNA products is achievable.

THE PROVESTEEN PROCESS - AN ULTRA-HIGH DENSITY FERMENTATION

The Provesta Corporation * pioneered the use of methanol as a fermentation substrate and has developed a number of bacteria and yeast methanol processes for the production of Single Cell Protein (SCP). One of these processes employs the continuous growth of a Pichia yeast on methanol at ultra-high cell densities and under conditions that lead to very high productivities. The fermenter effluent has a dry product solids content above 130 g/L that is dried directly after heat killing the yeast. Complex operations such as filtration and centrifugation are thus eliminated in the Provesta process, as are the difficult problems of maintaining sanitary conditions in such equipment. The elimination of major waste streams and the need to recycle

---

* Provesta Corporation is a wholly owned subsidiary of Phillips Petroleum Company.

spent medium by employing the direct dry process simplifies plant design, engineering and operations. The advantages of this improved ultra-high density direct dry process are listed in Table 1.

TABLE 1: Advantages of the Phillips-Provesta ultra-high density fermentation

1. Continuous Fermentation Process
2. Ultra-High Cell Density
3. Simplicity of Design
4. Maximum Use of Nutrients
5. Little Chance of Contamination
6. Many Substrates
7. Multiple Microorganisms
8. Elimination of Waste Streams
9. Environmental Protective
10. Modular Plant Construction

This technology has been expanded to include feedstocks other than methanol and includes such diverse substrates as glucose, sucrose, molasses, ethanol, whey, etc. In addition, the number and types of microorganisms which can be grown at such high cell densities in a continuous mode has also been expanded. This increases the versatility of this process technology and offers many opportunities for the practical production of biomass and products which previously were uneconomical or difficult to achieve. The potential of the Provesteen technology to grow a variety of yeasts at ultra-high cell density (125-160 g of dry cells per liter) can be demonstrated by data on the production of some Provesteen products.

The Phillips-Provesta process, as diagrammed in Figure 1, uses special fermentors designed to optimize oxygen and heat transfer. Oxygen transfer rates in excess of 800 mmol O_2/L•h are achieved in the process systems. Biofluid side heat transfer coefficients in excess of 1100 J/m^2•s•K can be obtained with specially designed tube baffle heat exchangers. In system operations, the medium composition and fermentation conditions for the chosen yeast and feedstock are tailored to give maximum yield of cells and maximum protein content. The carbohydrate substrate is used as the sole carbon and energy source and is the limiting nutrient throughout the fermentation. The concentration of dissolved oxygen is maintained at as low a level as possible while still having no oxygen limitation. The air flow rate is maintained at a level at which there is no CO_2 inhibition. Ammonia is supplied for pH control and as a nitrogen source. The system is controlled at the optimum growth temperature for the selected culture. Under this regime of continuous optimized growth conditions, the culture always remains in the most active log state of growth and maximum productivity is maintained.

The fermentation parameters for growth of *Torula utilis* on various carbon substrates utilizing this technology are shown in Table 2. The data include results that have been obtained using small laboratory size fermenters as well as pilot plant operations in a 1500 liter unit. The data show that productivities of above 20 g/L•h should be considered as achievable in fermentations for the production of biomass. Test results with other substrates and other yeasts show similar results to those reported.

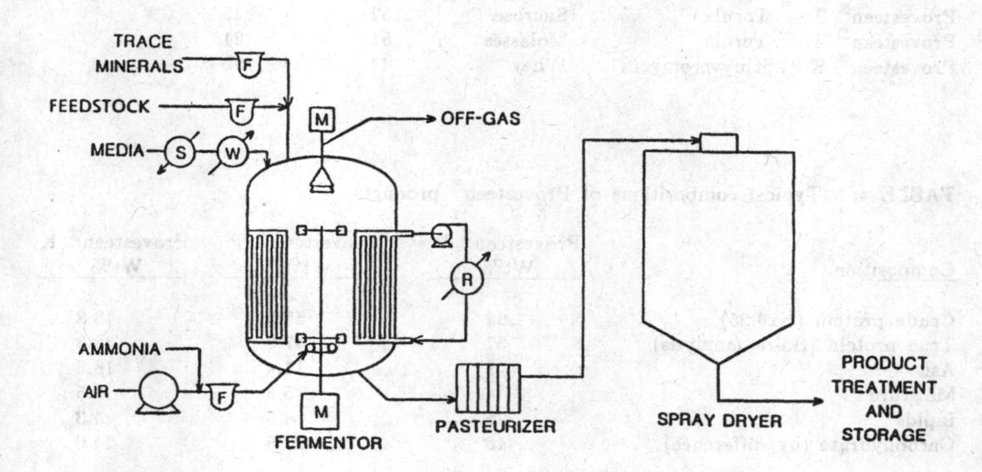

FIGURE 1: Phillips continuous Provesteen® single cell protein process

TABLE 2: Fermentation parameters of ultra-high density fermentations for production of Provesteen®T

Parameter	Sucrose	Ethanol	Molasses	
	Pilot Plant	Laboratory	Laboratory	Pilot Plant
Dilution Rate (h^{-1})	0.16	0.2	0.17	0.20
pH	3.9	3.8	4.5	4.5
Temperature (°C)	32-35	32	35	35
Fermentation Broth				
Total dry solids (g/L)	150	140	320	220
Washed cells (g/L)	135	127.4	160	105
Productivity (g/L•h)				
Based on total dry solids	25	28	55.9	44.1
Based on washed cells	23	25.4	27.1	21
Yield (g cells/g substrate)	0.52	0.8		
Oxygen Transfer (m mol O_2/L•h)	780	714	1100	870
Heat Transfer (Kcal/L•h)	92	70	126	100

TABLE 3: **Representative Phillips-Provesta yeast processes**

Product	Yeast	Feedstock	Yield, %	Productivity g/L•h
Provesteen® P	Pichia	Methanol	40	12
Provesteen® T	Torula	Ethanol	78	25
Provesteen® T	Torula	Sucrose	52	25
Provesteen® T	Torula	Molasses	51	21
Provesteen® K	Kluyveromyces	Whey	44	20

TABLE 4: **Typical compositions of Provesteen® products**

Composition	Provesteen® P Wt%	Provesteen® T Wt%	Provesteen® K Wt%
Crude protein (Nx6.25)	62	55.3	43.8
True protein (Buiret analysis)	57	47.0	38.3
Ash	1	11.4	18.8
Moisture	4	5.5	7.5
Lipids	5	4.6	4.3
Carbohydrate (by difference)	18	22.5	24.0
Mineral Content			
Calcium	0.09	0.07	1.0
Magnesium	0.3	0.35	0.3
Phosphorus	2.5	3.70	4.1
Potassium	2.5	1.80	4.9
Sodium	0.01	0.02	2.2
Trace Elements	ppm	ppm	ppm
Iron	340	271	352
Copper	36	36	38
Zinc	124	203	260
Manganese	26	19	23
Molybdenum	-	21	22

TABLE 5: Typical composition of Provesteen® T from sucrose

	Wt%		
Crude protein (Nx6.25)	57.0	Fatty Acids	% of Total
Ash	9.8	Palmitic	14.2
Crude fat	4.6	Palmitoleic	6.2
Moisture	5.5	Heptadecanoic	2.2
Crude fiber	<1.0	Stearic	1.0
Carbohydrate	22.5	Oleic	20.3
Minerals	Wt%	Linoleic	49.0
Potassium	2.75	Linolenic	5.6
Magnesium	0.34	Vitamin Content	Unit (mg/kg)
Sodium	0.02	Thiamine	9.5
Calcium	0.06	Riboflavin	44.1
Trace Minerals	ppm	Pyridoxine (HCl)	79.1
Iron	362	B_{12}	0.01
Zinc	252	Biotin	0.36
Copper	36	Choline chloride	4090
Molybdenum	24	Folic acid	21.5
Manganese	23	Inositol	3340
		p-Aminobenzoic acid	121
		Niacin	450
		p-Pantothenic acid	189

Three yeast products that are produced by the Phillips-Provesta high cell density process are Provesteen® P, T and K. As the name indicates, these products are produced with Pichia, Torula, or Kluyveromyces cultures. The Torula culture can use various substrates such as glucose, sucrose, molasses or ethanol. The Provesteen products are produced at high productivities (Table 3) while maintaining good yield. Comparisons of some typical compositions of these high cell density direct dried Provesteen products are shown in Table 4. The composition of the molasses grown cells are not shown due to the wide variation in the composition of molasses feedstocks, but the product generally is similar to other sugar grown cells. It should be noted that the Provesteen® P and T were designed to be very low in sodium content. This is desirable both for the product to be used in a low sodium human food supplement and it also decreases corrosion in the equipment. It also demonstrates that close control of the product composition is possible. A more extensive analysis of a typical composition of Provesteen® T produced using sucrose as the substrate is shown in Tables 5 and 6.

TABLE 6: Typical amino acid profile of Provesteen® T from Sucrose

Amino Acid	% W/W
Lysine	3.9
Histidine	1.1
Arginine	3.5
Aspartic acid	4.9
Threonine	2.8
Serine	3.0
Gutamic acid	9.2
Proline	2.0
Glycine	2.5
Alanine	3.5
Valine	3.0
Methionine	0.6
Isoleucine	2.5
Leucine	3.9
Tyrosine	1.9
Phenylalanine	2.4
Tryptophan	0.5
Cystine	0.2
Total	51.4

The Provesteen® products are fine powders that can be used as is or processed further into extruded material, pellets, granules or flakes. All animal feeding and digestibility tests indicate that Provesteen SCP is a high quality protein of nutritional value equal or superior to other sources of protein.

Besides the use of this technology to produce food grade yeast, the process can be used in the production of rDNA products in both yeast and bacterial cultures. Based on the results shown, this technology is truly an improved method for producing biomass. Additionally, this simplified process has applications in other fermentations and biotechnology-related processes. This technology should merit consideration by engineers who are involved in current and future fermentation operations.

33

MICROBIAL BIOMASS PROTEIN GROWN ON EDIBLE SUBSTRATES: THE INDIGENOUS FERMENTED FOODS

Dr. Keith H. Steinkraus
Professor of Microbiology
Institute of Food Science
Cornell University
Geneva, New York
U.S.A. 14456

ABSTRACT

Microbial biomass protein (MBP) production on inedible substrates is highly sophisticated, capital intensive processing and generally yields food/feed ingredients rather than foods. Further complex processing and formulation are generally required for human foods. Nevertheless, production of MBP for animal feeds could release immense quantities of cereal grains and legumes, presently fed to animals, for feeding humans, vast numbers of whom are hungry and starving. The indigenous fermented foods, MBP grown on edible substrates, generally require rather simple processing and low-cost factories. They yield foods directly, without further formulation, that are acceptable in flavour and texture to millions of people. The indigenous fermented food fermentations also can be used to produce meat flavours and textures, to enrich cereals, legumes and tubers with vitamins and essential amino acids, to raise the protein content of high starch substrates such as cassava and to decrease the cooking fuel requirements. The indigenous fermented food fermentations can be adapted to large, modern food processing as has been domonstrated by the Japanese shoyu, miso, natto, and sake fermentations and by the south African sorghum/maize beer and Mexican pulque fermentations. Feeding a human population of an estimated 8 billion people in the 21st century is going to require development, use and expansion of both MBP grown on inedible substrates and MBP grown on edible substrates-the indigenous fermented foods.

INTRODUCTION

The production of microbial biomass protein (MBP) on inedible substrates is one of the scientific miracles of the 20th century. The fastest growing plants require from 2 to 3 weeks to double their cell mass and considerably longer to produce the cereal grain and legume seeds that serve as the base for the human diet. Bacteria can double their cell mass in from 20 to 60 minutes while yeasts can double their cell mass in from 1 to 3 hours. Some molds can also double their cell mass in 3 to 6 hours. Thus, when the objective is to produce cell mass and protein by the most rapid and most efficient means, the bacteria, yeasts and molds offer the greatest promise of supplying the additional protein needed for both animals and humans as world population reaches 6 billion by the year 2000 and from 8 to 12 billion in the 21st century. Considering that, at the present time, approximately 1 billion of the world's people are hungry, prospects for the future include an even higher incidence of hunger, starvation and malnutrition. Producing foods and feeds from microbes will hopefully alleviate the problem.

All animal life on earth is dependent upon the photosynthetic plants, blue green algae and to a lesser extent photosynthetic bacteria for carbohydrates, lipids, proteins and all other food nutrients except perhaps vitamin B_{12} which appears to be synthesized only by prokaryotic bacteria. The plants have another advantage over animals as they can synthesize proteins using inorganic nitrogen and the legumes have still another advantage as they have nitrogen-fixing bacteria in their root nodules and can synthesize protein from atmospheric nitrogen. The only other sources of protein are the microbes which also can use inorganic nitrogen and those that are nitrogen fixing. Assuming that the total productivity of protein by plants will reach a maximum sometime in the future, man will have to depend upon the protein-producing capacity of microbes to make-up for any deficiencies.

At the present time, cereal grains and legumes, along with fish meal and dried yeast recovered from brewery and distillery operation, are prime ingredients for feeding poultry in the production of broilers and eggs, feeding swine in the production of pork and feeding other non-ruminant animals. They are used also for feed-lot fattening of beef even though the ruminants are the most efficient fermentors of lignocellulosic residues. Ruminants feeding on the range and utilizing lignocellulosic grasses are not consuming any proteins directly consumable by man and therefore, they can make a very valuable contribution to the foods and protein nutrition of the consumer. Animals such as chickens and pigs, consuming plant protein that could be consumed by man, are relatively inefficient in their conversion of plant to animal protein, such conversions ranging from 33% in the case of milk to lower efficiencies with eggs (27%), chicken broilers (18%), pork (9%), and beef (6%) (Pimentel et al., 1975). It is much more efficient protein usage to feed cereal grains, legumes and tubers to humans directly. This efficiency will be fostered when microbial biomass protein (MBP) becomes available on a large scale and it can be used to replace the cereal grains/legumes presently fed to animals.

Microbial biomass protein (MBP) can be produced on inedible substrates using inorganic nitrogen for protein synthesis. MBP compares well in its protein content compared with fishmeal and soybean meal (Table 1).

TABLE 1: Comparison of BP yeast grown on N-alkanes, and ICI bacteria grown on methanol with fishmeal and soybean meal proteins

	BP-Yeast	Fishmeal	Soybean Meal	ICI Bacteria
Crude Protein (N x 6.25)	66%	66%	48%	80%
True Protein (sum of anhydro amino acids)	49	50	38	58
Lipid	8	5-10	1.5	8
Ash	6	20	10	8

Fishmeal is richer in the sulfur-containing amino acids than MBP or soybean protein and is therefore a better protein than either MBP or plant protein for feeding animals or humans. However, by fortifying MBP or plant proteins with their limiting amino acids, proteins with excellent protein quality can be formulated. Combining proteins from two or more sources generally results in an improved protein value.

Production of MBP on inedible substrates is not without its problems. If the substrate is methanol, N-alkanes or even dilute starch, the MBP cells must be recovered from the substrate. Microbes grown on lignocellulosic waste generally will be consumed with the substrates. This limits the use of such MBP generally to ruminant animals capable of digesting the residual fiber. Other problems with MBP (Single Cell Protein - SCP) production include the following:

1. High capital cost for MBP factory. MBP (SCP) factories are generally very sophisticated involving the highest degree of microbial technology. Hence they are costly if they are designed to produce the optimum yield, namely 100,000 tons dried MBP/year. Such a factory will cost at least $75,000,000 and perhaps more. There are exceptions to these figures particularly in those cases when MBP is produced on waste lignocellulose or other food wastes such as citrus peels, pineapple wastes, etc. and the final product is consumed by cattle or other ruminants. Such factories can be much simpler and cheaper.
2. Requirement for large quantities of substrate at low cost. The technology is capital intensive. In order to produce MBP (SCP) at a low cost, the substrate must be cheap. Lignocellulosic substrates such as straw, sawdust, etc. are, in many cases, low cost but transport to a central processing factory may be costly. Some starches are relatively low cost. The yield of microbial cells can reach 1/2 gram per gram starch utilized. A factory producing 100,000 tons of MBP will require a low cost source of at least 200,000 tons of low cost starch/year. This involves not only transport of the starch to the factory, but substantial storage facilities as well.
3. Requirement for oxygenation in large fermentors. Microbes being grown at maximum efficiency require large volumes of oxygen supplied as efficiently as possible to the cells. This is costly in the design of large fermentors. Also, cells growing at their most rapid rates, produce considerable heat so that generally fermentors must also have considerable cooling capacity to maintain the fermentor temperature at optimum temperature levels.

4. Finally, microbial cells (MBP) harvested and separated from the substrate are generally not foods in themselves. Even the Rank, Hovis MacDougall (RHM) fungal mycelial protein, which comes close to being a food in itself, requires processing of the mycelial fibers to develop the meat-like texture and requires addition of fats and other flavouring materials to produce a food.
5. Rupture/fractionation of MBP cells. MBP (SCP) will make its greatest contribution to the world food problems when it can be fed directly to animals and humans as protein sources and the cells can be processed much as cereal grains/legumes are today. The technology is already available to rupture MBP cells, fractionate them, concentrate the proteins and either spin the proteins into fibers or to formulate and extrude the MBP to produce new foods resembling those presently manufactured from cereal grains/legumes and tubers. The Rank, Hovis MacDougall fungal mycelium does not require such extensive additional processing. Thus, it will likely be one of the first industrially successful MBP/SCP products to reach the commercial market in the Western World.
6. Functional properties of MBP proteins. If MBP/SCP is going to replace or extend proteins such as egg, soybean, cereal, etc. in food manufacturing, either in the food processing industry or in the homes, the functional properties include the following: flavour/blandness; colour/whiteness; solubility; viscosity; gelation; water uptake and retention; emulsifying capacity and ability to foam.
7. Nutritional value of MBP proteins. As with all proteins consumed as food, the nutritional value of MBP must be studied. Obviously, the nutritional value depends upon the type of organism, the species and strain, the conditions of growth, etc.
8. Toxicity of MBP proteins. Organisms vary in their toxicity and each MBP must be studied under a variety of growth conditions to insure that it never produces toxins against man or animals.
9. Flavor/texture acceptability of MBP proteins. Finally, MBP must have a flavour and texture acceptable to the consumer, whether animal or human, if it is to succeed in the market place.

EDIBLE MUSHROOMS GROWN ON LIGNOCELLULOSIC WASTE - A SPECIAL TYPE OF MBP

The most direct and the most acceptable form of MBP available for humans today in both the developed and developing world is mushroom which can be cultivated on lignocellulosic and other agricultural and food wastes. Edible mushrooms are foods in themselves highly prized by human consumers. They do not require capital intensive factories. They can be produced on any scale from household to large commerical production. There is no problem with functionality of the protein. Mushrooms are being extensively produced and consumed in Asia today and most people in the world, rich and poor alike, find them a very acceptable food. The subject of mushrooms as human food has been reviewed by Chang (1980). Fresh mushrooms of the Agaricus, Volvariella and Pleurotus types contain generally more than 3% protein, essential amino acid indices of 98 versus 99 for milk, amino acid scores of 89 versus 91 for milk and nutritional indices of 28 versus 25 for milk and 31 for soybeans.

One kilogram of dry composting material will yield as much as 1 kilogram of fresh mushrooms in 3 or 4 flushes over a period of 30 to 45 days. Usual yield is 600-750 grams of fresh mushrooms/kg dry compost. Considering that there are an estimated 2325 million tons of straw produced per year (FAO, 1977), over half of which may be burned, straw could be used as a substrate to produce approximately 1511 million tons of fresh mushrooms (65% efficiency) or 336 kg of fresh mushrooms annually for each of the present 4.5 billion human inhabitants of the earth (920 g fresh mushrooms containing about 28 g of protein per person/day). Microbial protein production on lignocellulosic and other wastes should be exploited to the fullest as an adjunct to the world's supply of food and protein. Following growth of the mushrooms, the spent beds can be used as a protein-enriched supplement for cattle and other animals, used as a nitrogen enriched soil conditioner or used as sources of cellulases and ligninases for the hydrolysis of other lignocellulosic substrates for the production of fermentation products.

MBP GROWN ON EDIBLE SUBSTRATES - THE INDIGENOUS FERMENTED Foods

The indigenous fermented foods also involve MBP but the substrates instead of being inedible are edible and involve the gamut of fresh vegetables, fruits/juices, cereals, legumes, tubers, milks, fish and other marine animals. The essential microorganisms include bacteria, yeasts and molds. In many cases the micro-organisms grow throughout the fermenting substrate. In certain cases, the microbial growth is mainly on the surfaces of the substrate. Generally, the major objective is not production of MBP as a predominate product in itself; rather the objective is to produce flavours and textures acceptable to the consumers. Fermentation therefore generally involves or includes production of sufficient enzymes to effect the desired degree of hydrolysis in the proteins, lipids and carbohydrates, yielding amino acids/peptide mixtures (meat-like flavours), organic acids and alcohols and other highly valued flavours and preservative compounds, synthesis of vitamins and essential amino acids and increasing the protein content of high starch substrates. Texture is produced by fungal mycelium in, for example, Indonesian tempeh or by precipitation of proteins and other components by acids as in the case of yogurts.

Advantages of indigenous fermented food protein over MBP grown on inedible substrates include the following:

1. The indigenous fermented foods are food (not food ingredients) already acceptable to millions of consumers.
2. They require low cost factories and processing.
3. They are adaptable to modern large scale processing (examples are Japanese shoyu, miso, natto and sake and South African sorghum/maize beer), all of which began as indigenous food fermentations that are large industries today.
4. The microbial protein is consumed with the substrate; no separation is required.
5. Oxygen is easily supplied.
6. Cooling capacity can be built into the process.
7. There is no requirement to rupture the cells to recover the proteins.
8. There is no problem with functionality of the proteins.
9. The indigenous fermented foods generally have decreased cooking fuel

requirements, quicker cooking times, better digestibility and improved nutritive value.

10. There are few toxic problems and when such problems do exist, they generally are well-known and can be avoided.

An interesting comparison is that of the Rank, Hovis, MacDougall fungal protein meat analogue, an example of MBP produced in large, sophisticated capital intensive fermentors and Indonesian tempeh kedelee (soybean tempe) produced by growing *Rhizopus oligosporus* or related molds on soaked, dehulled, partially cooked soybean cotyledons. Both products depend upon mold mycelium to provide the meat-like texture.

It is interesting to pursue the line of reasoning that may have led Rank, Hovis, MacDougall (RHM) to develop their fungal mycelia process (Spicer, 1971). They chose fungi because of the relatively low nucleic acid content compared with yeasts and bacteria and the fact that yeasts require supplementation with methionine to achieve satisfactory biological values. Fungal protein starts out with a net protein utilization (NPU) value equivalent to that of yeast protein fortified with methionine, about 70-75, while supplementation of fungal protein with 0.2% methionine increases its NPU to 100, equal to the protein quality of egg. Yeasts can replace a maximum of approximately 10% of the protein in a feed; fungal protein can be used up to 100% as a protein replacement. Fungal protein can be grown on low cost carbohydrates and its filamentous nature makes it easy to recover from submerged cultures by filtration. Fungi, particularly the mushrooms, are widely consumed and highly prized; and mold fermented foods such as Camembert, Stilton and Roquefort cheeses are highly acceptable to millions of consumers (Spicer, 1971). Consumption of fungi is also unaffected by religious taboos as pork and beef are. The mycelium, because of its fibrous structure, can be baked, fried or puffed to produce foods similar to those already consumed. Rank, Hovis, MacDougall was aware that the average Indonesian consumes 154 g of tempeh per day.

MEAT ANALOGUES (SUBSTITUTES)

It is interesting that it is the Western world, already the largest consumers of meats, that has devoted so much research and hundreds of millions of dollars to the development of meat analogues derived from wheat gluten, soybean protein and, most recently, fungal mycelium. All these processes are sophisticated and capital intensive. In the spun soybean process, soybean proteins are isolated and purified to about 92% purity, spun through platinum dies into chemical baths that yield fibrous strands much like hair. The protein fibers are then processed to yield meat-like hair. The protein fibers are then processed to yield meat-like textures, flavoured with meat flavourings and fats and yield synthetic bacon pieces, hamburger bits, synthetic ham and almost any type of meat-like product desired (Odell, 1966; Wanderstock, 1968). Swift and Company, a very large meat company, developed meat analogues using soybean protein concentrate which is formulated with the proper amounts of moisture, meat flavourings and fat and then extruded at very high pressures and temperatures to yield "chewey gel" nuggets that can be used as substitutes for meat in many foods. Yanschinski (1984) reports that Rank, Hovis MacDougall has joined forces with

Imperial Chemical Industries (ICI) already heavily involved in developing production of *Methylophilus methylotrophus* as a major MBP product called Pruteen. They are scaling-up production and test marketing the fungus *Fusarium graminearum* as a base for meat-like products. They are test marketing "fish cakes", "veal patties", and "meat-pies". Waste starch from the production of wheat gluten or hydrolyzed corn starch are used as substrates. The British Ministry of Agriculture and Fisheries has given the companies permission to produce enough product for market tests. If both scale-up and market tests are favorable, they may manufacture as much as 20,000 metric tons of fungal mycelium per annum.

To give the reader some idea of the expense of such research, ICI spent $150 million developing the Pruteen process. The RHM process is totally aseptic and can operate continuously over a period of time. The organism is grown at 30 ° C, pH 6.0 on glucose syrup (hydrolyzed corn starch), mineral salts, trace metals, choline and biotin. The amount of oxygen supplied to the fungus in the fermentor is very important. Too little results in anaerobioisis and development of undesirable by-products; too much oxygen inhibits growth. The product is a buff-coloured slurry with a slight mushroom aroma. The ICI technology which utilized an "airlift" fermentor has been modified to use more conventional mechanical stirrers. Immediately after growth, the fungus is killed by a thermal treatment that activates enzymes that breakdown ribonucleic acids which then pass out of the mycelium. The mycelium is harvested continuously by vacuum filtration on a horizontal belt filter where it forms a mat of interwoven mycelia bland in flavour and aroma and light in colour. The distribution of moisture in the filter cake is very important to the final texture. Colour, flavour and egg albumin are added to the mycelium and the mixture is steamed. The egg albumin sets the texture. Products requiring no special texture are simply molded to the desired shape. For meat substitutes such as chicken breasts, the fungal mycelium must be processed further by folding and refolding to align the fibers and develop true meat-like texture. The final products can be rehydrated in 15 seconds, do not shrink when cooked and can be canned or frozen with a 3 year shelf-life. RHM fungal protein contains 44% protein, 13% less than lean steak, 50% of the lipid in steak and no cholesterol and more fiber than whole-wheat bread. RHM has already produced over 50 tons of the products, most of which has been consumed in RHM canteens and restaurants (Yanchinski, 1984).

While the RHM fungal protein products are examples of very sophisticated microbial and food technology, they are a considerable contrast to Indonesian tempeh which serves as the meat analogue in Indonesia and part of Malaysia and is being adopted by the American vegetarian community.

The Indonesians, centuries ago, without modern chemistry and microbiology, developed a fermentation for making meat analogues from soybean in which mold mycelium provides the meat-like texture.

INDONESIAN TEMPEH KEDELE

Indonesian tempeh kedele is a white, mold-covered cake produced by fungal fermentation of dehulled, hydrated (soaked) and partially cooked soybean cotyledons. The mold grows throughout the bean mass knitting it into a compact cake that can be sliced thin and deep-fat fried or cut into chunks and used as a protein-rich meat substitute in soups. The essential molds are those belonging to the genus Rhizopus. *Rhizopus oligosporus* is the species identified as most characteristic and best adapted for production of tempeh (Steinkraus et al., 1960, 1983a; Hesseltine, 1961).

The essential steps in the productions of tempeh are the following:

1. Cleaning the soybeans
2. Hydration/acid fermentation
3. Dehulling dry or following hydration
4. Partial cooking
5. Draining, cooling, surface drying
6. Placing soybean cotyledons in suitable fermentation containers
7. Inoculating with tempeh mold (before or after placing in fermentation container)
8. Incubating until the cotyledons are completely covered with mold mycelium
9. Harvesting and selling
10. Cooking for consumption; deep fat frying or as an ingredient in soups in place of meat

TRADITIONAL TEMPEH FERMENTATION

The soybeans are washed and soaked in water overnight during which time the soybeans undergo bacterial acid fermentation reducing the pH to 5.0 or lower. An alternate process is to place the soybeans in water and bring it to a boil and then allow the beans to soak overnight. The general purpose of the boil is to facilitate hull removal. The hulls are removed by rubbing the soaked beans between the hands or by stamping them with the feet. The loosened hulls are then floated away with water. The cotyledons are then given a short boil, cooled, surface dried by winnowing and inoculated with tempeh mold, either from a previous batch of sound tempeh or from the mold grown and dried on leaves. Traditionally, the inoculated cotyledons are then wrapped in small packets using wilted banana or other large leaves and are incubated in a warm place for 2 or 3 days during which time the cotyledons are completely overgrown by the mold mycelium. The tempeh is then ready for cooking (Steinkraus et al., 1960).

INDUSTRIAL PRODUCTION OF TEMPEH

Twenty-five years ago most tempeh was prepared for sale using the traditional process described above. Then tempeh research in the United States resulted in some improvements in medium scale processing of tempeh. Steinkraus et al. (1965, 1983a) described a pilot plant process in which the soybeans were dehulled dry by passing them through a properly adjusted burr mill. Preceding the burr mill, the soybeans were given a short heat treatment at 104 °C (220 °F) to shrivel the cotyledons. The hulls were then removed from the cotyledons by passing them through an aspirator or over an Oliver gravity separator. Alternatively, the beans were soaked and dehulled wet by passing them through an abrasive vegetable peeler. Acidification of the beans, considered to be an essential step particularly in large scale processing where invasion by food-spoilage organisms could ruin large batches, was accomplished by adding 1% v/v lactic acid to the soak and cook water. The partially cooked beans were then drained, cooled and inoculated with powdered pure culture tempeh mold which had been grown on sterilized soybeans and freeze dried. The inoculum was mixed with the drained, cooled cotyledons in a Hobart Mixer. The inoculated beans were then spread on dryer trays (35 x 81 x 1.3 cm), covered with a layer of wax paper and incubated at 37 °C (98.6 °F) and 90% relative humidity. By this procedure, fermentation was complete in less than 24 hours. The tempeh was cut into 2.5 cm squares and the dryer trays were placed in a circulating hot air dryer at 104 °C (220 °F), dehydrated to less than 10% moisture, and packaged in polyethylene bags for distribution.

Within a few years, the commerical tempeh industry in Indonesia had adopted wooden trays with dimensions similar to those used above. They lined the trays with plastic sheeting perforated to allow access of air to the mold (Steinkraus, 1983a, 1983b; Shurtleff and Aoyagi, 1980; 1985).

Martinelli and Hesseltine (1964) developed a new method of incubating the tempeh in plastic bags with perforations at 0.25 to 1.3 cm intervals to allow access of oxygen. By this method the soybean cotyledons are inoculated with the mold and placed in the plastic bags or in plastic tubes similar to sausage casings. They can be incubated immediately or stored in a refrigerator until fermentation is desired. Then the mold overgrows the soybeans in a day or less. The plastic bag process has been widely adopted in Indonesia and is also being used commercially in new tempeh factories in the United States.

During the tempeh fermentation, not only is texture introduced into the soybean cotyledons, but also the proteins are partially hydrolyzed, the lipids are hydrolyzed to their constituent free fatty acids, stachyose, a tetrasaccharide undigestible in the human, is reduced, riboflavin nearly doubles, and niacin increases 7 times. Also, vitamin B-12, usually lacking in vegetarian foods, is synthesized by a bacterium that grows along with the essential mold (Liem et al., 1977). The bacterium has been identified as *Klebsiella pneumoniae* (non-pathogenic strain), Curtis et al., 1977. Thus, tempeh with its high protein content (about 40% dry basis) is not only able to supply the consumer with the essential protein requirements but also with the requirements for vitamin B-12.

The tempeh fermentation has been applied to a wide variety of bean types. (Gandjar, 1977). A new type of tempeh in which wheat and soybean are combined was developed by Wang and Hesseltine (1966; Hesseltine and Wang, 1979).

The tempeh process is a resource of considerable potential industrial value as it can be used to introduce meat-like textures into cereal/legume substrates. It also decreases the total cooking time required for soybeans from about 5 or 6 hours to about 40 minutes boiling (30 minute boil for a pre-cook and 10 minutes boiling following fermentation). It is one of the world's first quick-cooking foods - a quality highly prized in modern food science today.

Tempeh has been rapidly adopted by American vegetarians and it is becoming increasingly available in health food stores and even large supermarkets in both Eastern and Western United States. Tempeburgers are available on the West Coast.

According to Shurtleff and Aoyagi (1984), there are 53 tempeh factories in the United States, the largest of which produces (7,000 lbs.) (3182 Kg) of tempe/day; the largest operation in Indonesia produces (1760 lbs). (800 Kg) of tempe/day according to Shurtleff and Aoyagi (1980). Thus, tempeh production is still a relatively small commercial operation. Looking toward the future, if tempeh should become a staple in the diet of Americans, it is conceivably that tunnel fermentors would be used in which soybeans would be cleaned and dry dehulled, soaked and cooked, cooled, inoculated and passed continuously through a tunnel fermentor. The fermented tempeh would emerge from the tunnel fermentor following an approximate 20 hour processing time.

The Indonesian tempeh production of 154,000 metric tons/year (Table 2) dwarfs production in the rest of the world. Production of tempeh in the United States totaled 900 metric tons in 1983 (Shurtleff and Aoyagi, 1984) with a retail value of about $5,000,000. Based upon an estimated 10,000,000 vegetarians in the United States, only 90 grams of tempeh were available per vegetarian per year, a very small quantity. I do not know what the current price is for tempeh in Indonesia, but the prices both wholesale and retail in the United States are relatively high.

TABLE 2: Tempeh production in Indonesia (1976)[a]

Companies	41,201
Workers	128,000
Annual Production (metric tons)	153,895
Annual Retail Value ($US millions)	85.5
Utilization of Total Soybean Supply (%)	64

[a] - Winarno, 1976

There is a considerable contrast between the RHM fungal protein meat analogue and Indonesian tempe, both of which depend upon fungal mycelium for meat-like texture. The RHM product is nearly 100% MBP. The percent MBP in tempeh is likely in the range of 3 to 4%. The RHM product is capital intensive, highly sophisticated technology. Tempeh can be manufactured as a cottage industry or by modern food processing methods. The over-riding value of the tempeh fermentation is that it provides a low-cost method of introducing meat-like textures into cereal grain and legume substrates and also into food by-products such as peanut presscake, coconut presscake and soymilk/tofu residues that otherwise can be fed only to animals. The fermentation also provides a method of enriching cereal grains/legumes with vitamins at very low cost and decreasing cooking time. Vitamin B_{12} production in tempeh is a feature that should be emphasized as most vegetarian foods do not contain nutritionally signficant amounts of that vitamin.

RHM fungal protein and tempeh are not competitors. They actually serve two separate purposes and each can contribute much to feeding the world of the future.

This paper has concentrated on the fungal mycelial processes. Space does not permit detailed analysis of the contributions other indigenous fermented foods make to feeding the world. Another process well-worth mentioning is the Indonesian tape ketan/tape ketella fermentation (Steinkraus, 1983a; 1983b). In tape, *Amylomyces rouxii* and *Endomycopsis fibuliger* convert rice or cassava to sweet/sour alcoholic desserts. In doing so they can double the protein content, which is particularly important in areas subsisting on cassava.

The Chinese soy sauce/Japanese shoyu fermentations yield meat-like amino acid/peptide sauces of high nutritive value from vegetable proteins. Japanese miso converts rice or barley and soybean to meat-flavoured pastes. The lactic acid fermentations are responsible for preserving the nutritive value of vast amounts of vegetables around the world. The alcoholic fermentations produce flavours, and aromas that are highly appreciated by consumers and at the same time enrich the products with B-vitamins at very low cost. Although the level of MBP growth in many of these fermentations is rather low, on a total, overall world consumption basis, the MBP in the indigenous fermented foods has to be a very important contribution to our total food supply.

Both MBP grown on inedible substrates and MBP grown on edible substrates, the indigenous fermented foods, will be an important part of our feeds and foods of the future.

REFERENCES

*Curtis, P.R., R.E. Cullen, and K.H. Steinkraus. 1977. Identification of the bacterium producing vitamin B-12 activity in Indonesian tempeh. Symposium on Indigenous Fermented Foods (SIFF) Bangkok, Thailand. Nov. 21-27.

*Gandjar, I. 1977. Tempeh benguk; tempeh gembus; tempeh kecipir. Symposium on Indigenous Fermented Foods (SIFF), Bangkok, Thailand. Nov. 21-27.

Hesseltine, C.W. 1961. Research at Northern Regional Research Laboratory on fermented foods. In: Proceedings of Conference on Soybean Products for Protein in Human Foods. pp. 67-74. USDA, Peoria, IL. September 13-15.

Hesseltine, C.W. 1965. A millennium of fungi food and fermentation. Mycologia LVII: 149-197.

Hesseltine, C.W. and H.L. Wang. 1979. Fermented Foods. Chem. & Industry June: 393-399.

Liem, I.T.H., K.H. Steinkraus, and T.C. Cronk. 1977. Production of vitamin B-12 in tempeh, a fermented soybean food. Appl. & Environ Microbiol. *34*: 773-776.

Martinelli, A. and C.W. Hesseltine. 1964. Tempeh fermentation: Package and tray fermentations. Food Technol. *18*: 167-171.

Odell, A.D. 1968. Incorporation of protein isolates into meat analogues. In: Singel-Cell Protein edited by R.I. Mateles and S.R. Tannenbaum. M.I.T. Press, Cambridge, MA.

Pimentel, D., W. Dritschilo, J. Kummel and J. Kutzman. 1975. Energy and land constraints in food protein production. Science. *190*: 754-761.

Shurtleff, W. and A. Aoyagi. 1980. Tempeh Production. New Age Foods.

Shurtleff, W. and A. Aoyagi. 1984. Soy Foods Industry annd Market 1984-85. Soyfoods Center, Lafayette, CA. 94549.

Shurtleff, W. and A. Aoyagi. 1985. The Book of Tempeh. 2nd Edition. Harper and Row.

Skinner, J.J. 1975. Single-cell protein moves toward market. Chem. & Eng. News. *53* (18): 24-26.

Spicer, A. 1971. Synthetic proteins for human and animal consumption. Vet. Record. *89*: 482-486.

Stahel, G. 1946. Foods from fermented soybeans as prepared in the Netherlands Indies. II. Tempe, a Tropical staple. J. N.Y. Bot. Gardens. *47*: 285-296.

Steinkraus, K.H. (Editor) 1983a. Handbook of Indigenous Fermented Foods. Marcel Dekker (New York).

Steinkraus, K.H. 1983b. Traditional food fermentations as industrial resources. Acta Biotechnologica. *3*: 1-12.

Steinkraus, K.H., Y.B. Hwa, J.P. Van Buren, M.I. Provvidenti, and D. B. Hand. 1960. Studies on tempeh - An Indonesian fermented soybean food. Food Res. *25*: 777-788.

Steinkraus, K.H., J.P. Van Buren, L.R. Hackler, annd D.B. Hand. 1965. A pilot-plant process for the production of dehydrated tempeh. Food Tech. *19*: 63-68.

Wanderstock, J.J. 1968. Food Analogs. Cornell H.R.A. Quarterly. pp. 22-33. August.

Wang, H.L. and C.W. Hesseltine. 1966. Wheat tempeh. Cer. Chem. *43*: 563-570.

Winarno, F.G. 1976. The Present Status of Soybean in Indonesia. FATAMETA. Bogor Agricultural Univesity.

Yanchinski, S. 1984. U.K. sinks its teeth into myco-protein. Biotech. November: 933.

* Summaries of these papers are included in Steinkraus (1983a).

TECHNO-ECONOMIC EVALUATIONS OF VARIOUS SUBSTRATES FOR SCP PRODUCTION: A CASE STUDY FOR MEXICO

M. de la Torre-Louis and L.B. Flores Cotera
Department of Biotechnology and Bioengineering
DINVESTAV-IPN,
Mexico, D.F.
Mexico

SUMMARY

Taking into consideration bench scale data of SCP production from sugar cane bagasse (pith), cassava and available commercial production data for SCP from methanol and molasses, the economics of each of these substrate alternatives are compared. It is concluded that the economics depend strongly on the carbon source price. The selection of substrate depends on price and availability of raw materials at each location. With regard to capital investment and protein cost, SCP from molasses seems to be the best alternative for Mexico under current conditions.

INTRODUCTION

Single Cell Proteins (SCP) have been the subject of considerable technical effort over the world but in most instances, the economic and technological constraints have limited their industrial application. Although several SCP processes have achieved technical feasibility, very few developments have reached a commercial or even demonstration stage (1).

It is well known that fodder yeast production started in Europe at the beginning of the century and has continued since then with fluctuating success. However, this industry based mainly on the utilization of agricultural and industrial wastes (or byproducts) such as molasses and sulfite liquors never provided a significant fraction of the total protein supply (2). On the basis of protein shortage in the world, and considering technology evolved during the last two decades, there is reason to be optimistic about the future economic feasibility of SCP.

One of the major problems in Mexico is the shortage of proteins for human and animal nutrition. The potential of SCP as a supplementary source of proteins has encouraged a number of research groups in Mexico (3, 4, 5) to examine and develop processes for SCP production directed to animal feeds. A wide variety of raw materials has been considered, such as lignocellulosic residues, molasses, methanol, starchy substrates, n-paraffins and others. In industrial operations such as SCP production, it is often possible to produce equivalent products in different ways. Although the results may be approximately the same, the capital required and the expenses involved can vary considerably, depending on the raw materials and processes chosen. It is necessary, therefore, not only to decide if SCP production would be profitable, but also, to decide which of several possible raw materials would be the most desirable.

In this work, considering bench scale data obtained in our institute for SCP production from sugar cane bagasse and cassava, and available industrial data of SCP production from methanol (England) and molasses (Cuba), the capital investment and the total product cost are compared for processes utilizing the above mentioned substrates. The fermentation facility is designed assuming a production capacity of 50,000 tons/yr, operating, 300 days/yr and using the design considerations given in Table 1. Equipment costs are estimated from the figures of references 12, 13, 14, 15 and 16 or from direct manufacturers quotations and up dated to first quarter 1985, using the Marshall and Swift Equipment Cost Index (17).

TABLE 1: Comparison of the four SCP processes considered (in US dollars)

	PROCESS			
	Bagasse	Molasses	Methanol	Cassava*
Investment ($X 10^6$ $)	78.204	48.770	77.200	--
Unit Product Cost ($/Ton)	1098	756	1264	605[a]
Raw Materials and Utilities ($/Ton)	465	332	618	212[b]
Protein Content (%)	60	52	71	17
Protein Cost ($/Ton)	1830	1454	1780	3558

* - Preliminary results only

a - Raw materials assumed at 35% of total unit product cost.

b - Based on raw materials only (utilities not included)

SCP PRODUCTION FROM SUGAR CANE BAGASSE (PITH)

Sugar cane bagasse is one of the most promising substrates for single cell protein production because it is a byproduct generated from the processing of a renewable resource, its production is localized and it is relatively cheap. Currently, the sugar cane industry in Mexico produces over 11 million tons of bagasse annually which has been mainly used as a fuel and to a lesser extent for paper and board production. In the future, it could be more desirable to use bagasse as substrate for SCP production and contribute in this way, to solving one of the most demanding problems in Mexico: protein shortage for animal nutrition.

A process for SCP production from sugar cane bagasse using a mixed bacterial culture of *Cellulomonas flavigena* and *Xanthomonas sp.* (6, 7, 8, 9) has been developed in our institute (Figure 1). Based on bench scale and 14 and 70-L fermentor data, the main technical problems are identified and the economics of the processes are analyzed.

In this process, one part of bagasse is treated with ten parts of water containing 2% w/w NaOH at 93°C for about 15 minutes. To make efficient use of NaOH, the supernatant fractions are recycled for the treatment of several batches of bagasse (up to six) after re-adjusting the alkali concentration to 2%. For this stage, we have designed and considered for capital estimate purposes, a continuous counter-current treatment unit. The mixed bacterial culture of *Cellulomonas flavigena* and *Xanthomonas sp.* is grown on a medium containing treated bagasse and mineral salts. The system operates as a cyclic fed batch culture, within an agitated fermentor. After each fermentation cycle, 11.5% by volume of the fermented broth is left behind in the fermentor, in order to keep a high cell concentration at the beginning of the succeeding batch. This high initial cell concentration and the treatment conditions of bagasse permit the direct feeding of bagasse to the fermentor without including a sterilization step (10, 11). The remainder of the fermented broth containing about 42.5 g/L of bacterial cells and residual substrate is removed from the fermentor and sent to the recovery section.

Residual substrate is separated from the bacteria cell cream on a rotary vacuum filter, dried and sent to storage. On the other hand, the biomass suspension is flocculated and preconcentrated by centrifugation to a dry biomass content of 14%. Before drying and in order to assure a better quality of the final product, the preconcentrated cream is thermolyzed and concentrated to 24% solids content in a falling film evaporator. The final drying of the cells is carried out on a direct contact spray dryer. The dry powder is separated from the drying gases in a battery of cyclones and packed in 25-Kg bags. Since some product losses occur during cell recovery in centrifugues and cyclones, a 98% recovery efficiency was considered.

Two products are obtained: A forage containing 12 to 15% protein and the microbial biomass with 60 to 65% protein. In regard to protein content and digestibility (60%), the forage is similar to dry alfalfa. The biomass in preliminary feeding trials on rats has shown good protein nutritional value comparable with soymeal.

Total captial investment is $78.20 million U.S. After taking a credit of $40.00 U.S./ton for forage byproduct, $1,098.00 U.S./ton must be charged for the SCP to provide a 15% after tax simple return on investment as profit. Preliminary results obtained at CENIC (Cuba) have shown (18) that caustic soda consumption can be considerably reduced, when treatment is carried out in solid phase, by spraying a high concentration NaOH solution over bagasse under controlled humidity conditions. In this way, treatment occurs over a period of days during bagasse storage. The advantages of solid phase treatment besides the lower chemicals consumption are: no steam requirements, lower capital investment, no waste produced and, since alkali soluble carbohydrates are not washed out, a higher substrate fraction is available for the microorganisms. The data indicate that it might be possible to obtain a 50 to

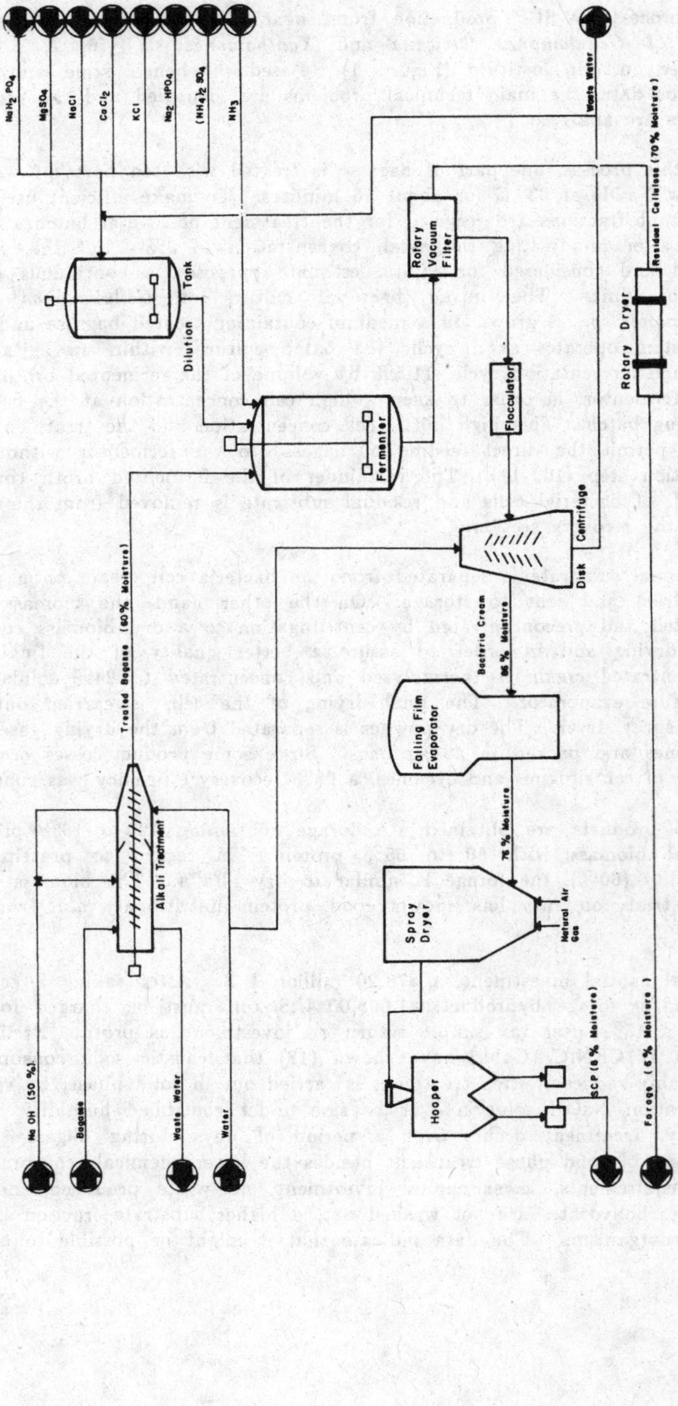

FIGURE I SCP PRODUCTION FROM SUGAR CANE BAGASSE

80% reduction in the alkali consumption as compared to the figures used for this analysis.

Mass doubling time is another very important factor, due to its effect on the volumetric productivity. It is obvious that at smaller doubling times, more biomass is being produced per unit time and less fermentors will be required to produce a given amount of SCP. In this work, a doubling time of 5.8 hr was assumed, but doubling times as low as 4.5 hr have been observed in our laboratories. Considering this later value, capital investment is reduced by about 10%.

Our results to date appear promising if we consider that by reducing alkali consumption to 60% of the actual value and capital investment by 10%, the SCP total product cost would be under $1,000 U.S./ton. Finally by recovering the solubilized hemicelluloses and lignin, the overall economy of the process could be improved, although it has not been considered here.

SCP PRODUCTION FROM MOLASSES

In 1984, the Mexican sugar cane industry produced 1,234,222 tons of molasses, 799,553 tons satisfied the domestic demand and 187,721 tons were exported. According to governmental estimates sugar production will increase in the near future, but the demand for molasses will not increase at the same level (19). On the other hand, since international molasses prices are currently low and probably will be even lower in the near future, its use as raw material for the national industries seems to be more desirable than of exporting it.

A flow diagram for fodder yeast production from molasses is shown on Figure 2. The main equipment specifications and raw materials and utilities consumption are based on industrial data available from 40 tons/day fodder yeast plants actually operating in Cuba (20).

Prior to the fermentation step, molasses should be treated to eliminate impurities, volatile organic acids and unwanted microorganisms. The molasses is diluted, heated to 80 °C and solids flocculated by lowering pH with sulfuric acid. Finally, the solids are separated by centrifugation and the molasses is sterilized in a continuous plate heat exchanger. A *Candida utilis* strain is grown continuously on a medium containing sterile molasses and mineral salts that are fed as independent streams to the loop stirred fermentor employed in this process. The fermented broth containing about 30 g/L dry weight of yeast is continuously removed from the fermentor and sent to the cell recovery section. The fermented broth is pumped to a demulsifier where air disengagemennt occurs, allowing a high cell recovery efficiency by centrifugation. The cream obtained is washed, and in order to assure a better quality of the final product, it is passed through a system integrated by preheater, thermolizer and falling film evaporator, which delivers a cream containing between 22 to 24% solids to the drying system. The drying of the cells is carried out in a direct contact spray dryer. The dry powder is separated from the drying gases in a battery of cyclones and packed in 25 kg bags. The product contains between 45 to 53% protein and high levels of vitamins, particularly the B Complex. It has been used for

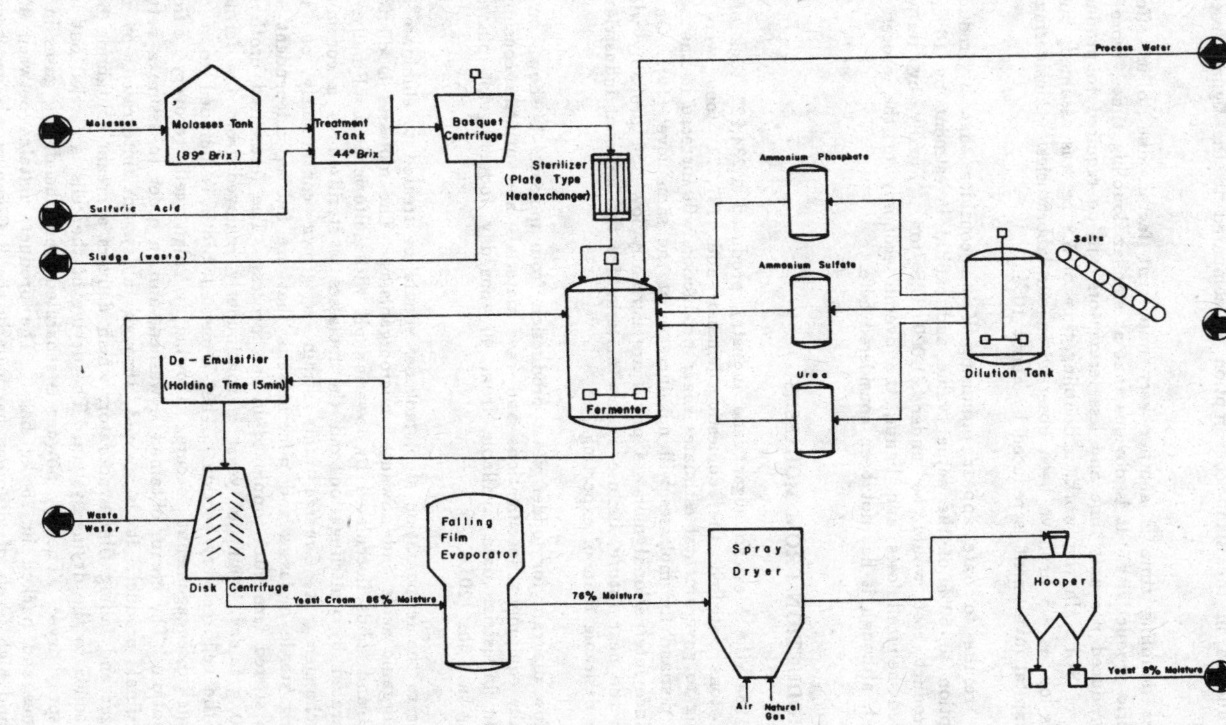

FIGURE 2 SCP PRODUCTION FROM MOLASSES

the last twenty years in poultry, pigs and calves feeds.

Total capital investment is $48.77 million U.S. After charging a 15% after tax simple return on investment as profit, $756 U.S./ton must be charged for the fodder yeast produced. In Mexico, the price of molasses depends on its final use. For animal feeding, molasses cost is $27 U.S./ton, while for other industrial purposes; e.g. ethanol production, the cost goes up to $83 U.S./ton. For the puposes of this work the first price was considered, but considering the highest price, $997 U.S./ton must be charged for the fodder yeast produced.

SCP PRODUCTION FROM METHANOL

According to official sources (21) a deficit of methanol in Mexico is expected by 1986, and consequently, substrate for SCP production will not be available until a new 825,000 tons/yr facility comes on stream by 1990. The LANFI Laboratories (a government dependency in Mexico City) studied several processes for SCP production from methanol (22). Based on a technical proposal from John Brown Engineers and Constructors Ltd. (UK) which built the ICI 60,000 ton/yr SCP plant in Billingham, they concluded that the ICI process was the most appropriate for Mexico. For the purposes of this work, the data from LANFI's technical report were adapted in order to put them on the same basis of comparison as the other processes considered.

A flow diagram of the ICI process is shown in Figure 3 (23, 24). Air, methanol, ammonia, and inorganic salts are sterilized and fed continuously to a single large pressure cycle fermentor, which has no moving parts and was invented by ICI. The micro-organism, a strain of the species *Methylophilus methylotrophus,* is maintained at a concentration of 30 g/L in the fermentor by continously withdrawing the circulating culture. The material from the fermentor passes through flocculation and separation stages to remove the bulk of the water which is then recycled to the fermentor. After separation of the bulk of the water, the product stream is then dried in a rotary dryer (or flash dryer) to produce a granular product which is used, typically in poultry feeds. Some of this material is ground to a fine powder to produce a product with the desired suspension properties for use in milk replacer diets for calves. "Pruteen", the trade name, is one of the most concentrated, highly digestible protein sources available for animal feeds. It contains 73% protein, 8.6% total lipids, with an amino acid profile high in lysine and methionine.

The total investment is $77.2 million U.S. After charging a 15% after tax simple return on investment, the total product cost is $1,258 U.S./ton. One of the main factors influencing the product cost on this alternative is the methanol price. A subsidized price was considered in this work although the price of methanol in the U.S.A. may be substantially higher. For this reason, SCP production from methanol seems to be advisable only for countries with low cost hydrocarbon sources available.

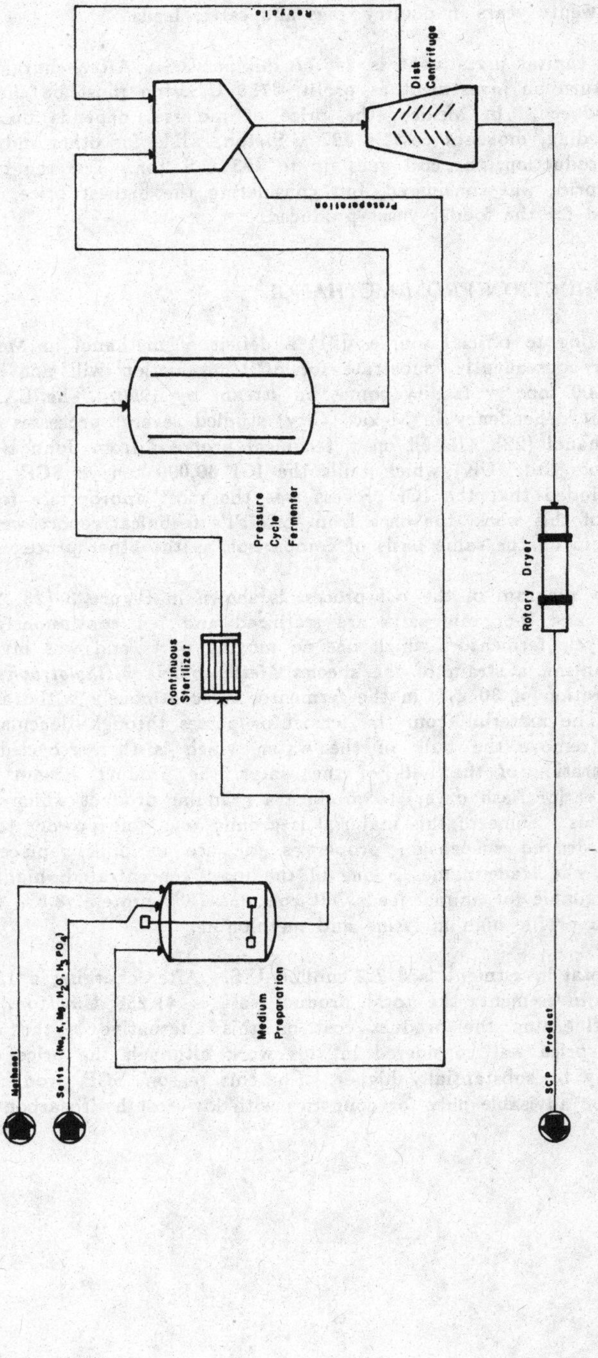

FIGURE 3 SCP PRODUCTION FROM METHANOL

Methanol

Salts (Na, K, Mg, H_2O, H_3PO_4)

Medium Preparation

Continuous Sterilizer

Pressure Cycle Fermenter

Precipitation

Recycle

Disk Centrifuge

Rotary Dryer

SCP Product

PROTEIN ENRICHMENT OF CASSAVA

Cassava roots have a very low protein content (1-2% DM) and, for that reason, it is desirable to upgrade it by fermentation in order to increase its value for animal nutrition. A process for protein enrichment of cassava has been developed at bench scale in our institution (25, 26, 27). The main stages of the solid state fermentation of cassava are shown in Figure 4. *Rhizopus oligosporus* NRRL 2710 is grown on a non-sterile medium containing cassava and mineral salts in a cabinet (0.4 x 0.35 x 0.5 m) equipped with three trays (0.3 x 0.3 m). After 48 hours of fermentation, the product contains between 17 to 20% true protein with an amino acid profile high in lysine.

At present, there is not enough experimental data to establish a complete process analysis. For that reason, only the raw materials cost per ton of product is estimated. However, unquestionably more research will be necessary before dismissal of cassava as raw material for SCP production occurs.

CONCLUDING REMARKS

Costs have been estimated for the production of SCP from four different substrates. It is apparent from our estimates, summarized in Tables 1 and 2, that raw materials account for 27 to 35% of the total product cost, utilities 10 to 18% while depreciation, profits and taxes combined represent 43 to 50% of the total product cost. In general, an equal split between capital and operating costs are observed in all the processes considered. Price and availability of raw materials play important roles in the process economics; therefore, the selection of process depends strongly on location. Capital investment and protein costs suggest that SCP production from molasses could be the first choice in Mexico. However, since soybean imports averaged 800,000 tons/yr recently, it is clear that in addition to molasses, other substrates must be considered to cope with the current national deficit of proteins.

TABLE 2: Components as percent of total product cost

Substrate	Raw Materials	Utilities	Depreciation	Profits and Income Tax	Other
Bagasse	32.5	9.8	10.5	39.9	7.3
Molassses	27.5	16.4	9.8	36.1	10.2
Methanol	31.0	18.2	9.3	33.8	7.7
Cassava	35.0	-	-	-	-

Methanol and sugar cane bagasse as primary raw materials for SCP production may be good alternatives; however, the main disadvantage of the current technologies is the high capital investment required. Substantial progress in the technology of SCP

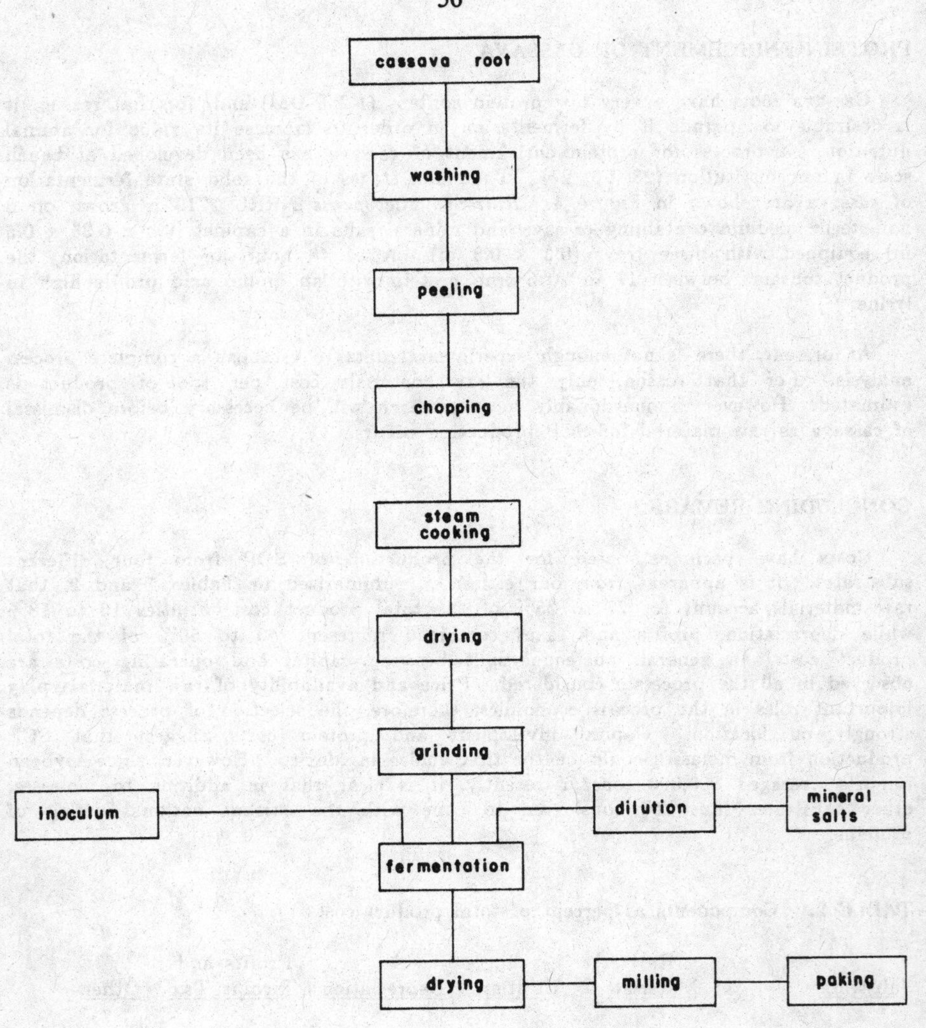

Fig. 4 Diagram for solid substrate fermentation of cassava

production is expected in the near future, mainly regarding to new strains developed by genetic engineering, fermentor design, and operations for cell separation and drying. The progress in these areas no doubt will contribute strongly to overcoming the main economic constraints for SCP production from these raw materials.

Finally, production cost obviously may not be the sole consideration when considering prospects for SCP. For a nation such as Mexico, importing large amounts of protein, it may be more important to overcome the corresponding hard currency expenditures to preserve its economic independence and to upgrade unutilized agricultural and natural resources.

ACKNOWLEDGEMENTS

The authors are indebted to Dr. A. Enriquez, CENIC (Cuba), for his advice and comments on the work. This work was carried out with the support of the Subsecretaria de Educacion e Investigaciones Tecnologicas, SEP/MEXICO.

REFERENCES

1) Anon. "Breathing new life into single cell protein" Chem. Eng. 91: (3) 19 (1984).

2) J.C. Senez "International Symposium on SCP", January 28, 1983. Paris, France.

3) De la Torre M., Contreras I., Flores L.B., Ponce M.T., Vazquez F. "Produccion de PUC y forrajes a partir de medula de bagazo de cana". (Report BBF-8501). CIEA-IPN, February 1985.

4) De la Torre M., Contreras I., Flores L.B., Ponce M.T., Vazquez F. "Produccion de Levadura Forrajera a partir de melazas". (Report BBF-8502). CIEA-IPN, February, 1985.

5) "La Investigacion en Biotecnologia y Bioingenieria". Coleccion Inventarios SEP/COSNET. Subsecretaria de Educacion e Investigacion Tecnologicas, Mexico (1984).

6) De la Torre M., Casas C. "Isolation and Characterization of a symbiotic cellulolytic mixed bacterial culture". Appl. Microbiol. Biotech. 19: (6), 430 (1984).

7) De la Torre M. "Produccion de Proteinas Alimenticias de origen unicelular en residuos lignocelulosicos". Ph.D. Thesis. Escuela Nacional de Ciencias Biologicas (1981). IPN, Mexico.

8) Lorencez I. "Estudios sobre la degradacion de celulosa por un cultivo mixto de bacterias". Thesis Escuela Nacional de Ciencias Biologicas (1979), IPN, Mexico

9) De la Torre M., "SCP Production from cellulosic wastes". "Conservation and Recycling". 5:(1) 41 (1982).

10) Callihan, C.D. Dunlap C.E. "Single cell protein from waste cellulose". Report PB 223873 National Technical Information. Service U.S. Departament of Commerce.

11) Han, Y.W., C.D. Callihan and H.A. Schuyten. "Combined effect of heat and alkali in sterilizing sugar cane bagasse". J. Food Sc. 36 (1971).

12) Hall R.S., J. Matley, K.J. McNaughton. "Current cost of Process Equipment". Chem. Eng. 89: (7) 80 (1982).

13) Perry, R.C. Green "Chemical Engineers Handbook". Sixth Edition, McGraw Hill (1984). New York, U.S.A.

14) Peters, M.S., K.D. Timmerhaus. "Plant Design and Economics for Chemical Engineers". Third Edition McGraw Hill (1980) New York, U.S.A.

15) Mulet, A., A.B. Corripio, L.B. Evans. "Estimate cost of pressure vessels via correlations". Chem. Eng. 88: (20) 145 (1981).

16) Mulet, A., A.B. Corripio, L.B. Evans, "Estimate cost of heat exchangers and storage tanks via correlations". Chem. Eng. 89: (2) 125 (1981).

17) Matley, J., "CE Cost indexes set slower pace". Chem. Eng. 92: (8) 75 (1985).

18) Enriquez A., CENIC, Cuba. Personal communication.

19) Informacion por la Gerencia de Desarrollo Industrial de Azucar, S.A. Mexico City, Mexico.

20) Estevez, R. "Levadura forrajera a partir de las mieles finales de cana". Los Derivados de la Cana. ICIDSA (Cuba), Editorial Cientifica (1980). La Habana, CUBA.

21) PEMEX. "Memoria de Labores 1984". Mexico City, Mexico.

22) LANFI. Proyecto Proteina Unicelular: Estidio Tecnico economico preliminar. Mayo 1984. Mexico City, Mexico.

23) Riviere, J. "Industrial Applications of Microbiology". John Wiley and Sons, New York, U.S.A. (1977).

24) Anon. "ICI Protein Process" Hydrocarbon Processing 62: (11) 147 (1983).

25) Ramos-Valdivia, A.C. "Enriquecimiento proteico de la Yuca (Manihot esculenta Crantz)", M.Sc. Thesis Centro de Investigacion y de Estudios Avanzados, IPN Mexico (1982).

26) Ramos-Valdivia, A., M. de la Torre and C. Casas-Campillo. "Solid State Fermentation of Cassava with *R. oligosporus* NRRL 2710". In Production and Feeding of SCP. Feranti and Fiechter Editor. Appl. Science Publishers. LTD. Great Britain (1983) pg. 104.

27) Raimbault, M. and D. Alazard. 1980. "Culture method to study fungal growth in solid fermentation". European J. Appl. Microbiol. Biotech. 9:199 (1980).

ECONOMICS OF SCP BASED ON LIGNOCELLULOSIC WASTES: A CASE STUDY FOR CANADA

N. Kosaric
Chemical and Biochemical Engineering

Peter C. Bell
School of Administration
University of Western Ontario
London, Canada N6A 3K7

ABSTRACT

This paper discusses some issues in the economic analysis of processes for the production of non-conventional microbial foods. One process, the Czechoslovak Academy of Sciences (Prague) process for the production of SCP from waste sulphite liquor from a paper mill, is examined in detail and the conditions which might make this process profitable in Canada are reviewed. Finally, some general conclusions with respect to process economics in Canada are presented.

INTRODUCTION

In order for non-conventional microbial foods to become a major factor in the Canadian marketplace, these products must be shown to be safe to consumers, technologically feasible to manufacture in quantity, and, equally importantly, economically viable for the producer.

This paper is based on unpublished work by Kosaric et al. (1984) and concentrates on the economics issues surrounding SCP manufacture. Following a discussion of the general factors which make economic analyses of SCP manufacture difficult or controversial, an analysis of one specific production process is presented. The process chosen for detailed analysis is a process developed by the Czechoslavak Academy of Science, Prague to produce 25,000 to 28,000 tons of an SCP product annually. This product which contains 45-50% protein is produced through fermentation of the waste sulphite liquor from a paper mill. The paper concludes with some general comments concerning Canada as a site for processes of this type.

ECONOMIC AND MARKETING ISSUES

There are four difficult, but critical issues that complicate economic analyses designed to investigate the viability of SCP manufacture. These are:

1. How to determine the product price?
2. How to handle the capital cost of building the plant?
3. How to assess the impact of inflation?
4. How to assess the impact of the tax system?

In most cases, the principal source of income to support manufacturing operations is the revenue from the sale of product. Since manufacturing facilities are designed for a specific scale of operation, the principal uncertainty in the sales revenue is the price/unit obtainable for the product. Estimation of a reasonable market price is both critical and difficult. The starting point for the process of price estimation is the marketing plan.

Bringing a new SCP product to market presents many of the same challenges as the introduction of any new product, and the key marketing decisions are the same. The correct time to commercialize the product, the appropriate target market, pricing, method of distribution, message and target of an advertising effort are all important elements of the marketing plan. However, the marketing of single cell protein presents some unique problems which increase the riskiness of the marketing effort. One particular problem is the difficulty in selling a cellular protein product to a consumer who psychologically associates bacteria, fungi, and to some extent, yeast, with disease. The consumer could be a farmer intending to feed SCP-containing meal to his livestock or the intended end-user of a SCP product designed for human consumption. A second source of consumer resistance is the fear of toxic carry-over from the substrate used for the fermentation. This is especially a problem with hydrocarbon-based processes but can also be a factor in carbohydrate and cellulose-based processes where the substrate could be in a toxic form initially (eg. some pulp effluents) or where toxic materials may be required to treat the substrate during, or prior to, fermentation (e.g. some pre-treatment methods of cellulosics).

The reluctance of consumers to eat SCP products, directly or indirectly, especially those derived from a toxic substrate, introduces considerable uncertainty into the problem of estimating an attainable market price. For this reason, it appears likely that the first SCP products will be marketed as protein supplements for animal feeds.

A SCP product marketed as a protein supplement for animal feeds will have to compete in the market with materials currently used as protein sources in feed formulations. These include soyabean meal, rapeseed meal, meat meal, feather meal, corn gluten, linseed, and others. Each available protein source has different nutrient levels and different physical characteristics which may limit the amount used in a feed.

The major "players" in the animal feed market are the feed formulators. Feed formulators produce complete feeds which are sold to farmers, either directly or through farm supply companies. Protein supplements are a key ingredient in most feeds. The feed formulators obtain the various protein sources from "processors" who process the raw materials into usable feed ingredients. The processors maintain stocks of finished product to meet the orders from the formulator. While the processors are often independent, several of the large feed formulators are now vertically integrated and control their own processing operations to ensure a supply of feed ingredients.

In 1980 there were 609 feed companies in Canada (excluding small business operations). Ontario and Quebec comprise a large percentage of the total Canadian operations with 196 and 226 feed operations, respectively. The total value of shipments of own manufacture from feed companies nationally, was $2,433 million in 1980 and $2,281 million in 1981, with Ontario and Quebec together accounting for $1,631 million of the 1980 total (Statistics Canada, 1981, 1983).

Ingredient costs are very important to the formulator. A breakdown of production costs (Statistics Canada, 1980) shows that 78 percent of production costs in the feed industry were raw material costs. The cost to the feed industry of raw materials and supplies in 1980 totalled $1,844.3 million, $1,316.6 million of which represents costs to feed companies in Ontario and Quebec. Of this national total, more than $375 million is accounted for by soyabean flour, cake, and meal, gluten meal, meat meal, feather meal, and rapeseed meal, all commonly used protein sources. Most of the materials used as protein supplements are imported unprocessed from the United States, and processed in Canada.

The largest single source of protein used in the industry are soyabean products. Imports of unprocessed soya beans in 1982 were 461,784 tonnes with a value of $128 million, while imports of processed forms of soya bean totalled 387,681 tonnes with a value of $108 million (Statistics Canada, 1982). Feed is generally formulated in and for domestic markets. Figures for bulk feeds, as opposed to feed ingredients, show imports of $15 million and exports of total value $17 million.

The decision to use a particular protein source in a particular feed meal is a complex one. The formulator must meet nutrient content standards set by the government and is constrained by the physical nature of the available materials (e.g. consistency), palatability, presence of growth-depressing and growth-promoting factors, and the availability of the protein sources. Finally, the formulator makes a profit by formulating his feeds so that they are cost competitive with feeds from other formulators that have the same nutrient content; thus, the formulator seeks minimum cost formulae that meet the quality, content and availability restrictions. The least cost formulation which will satisfy the myriad of constraints placed on the formulator is found using linear programming, a computerized optimizing technique. The constraints and the formulator's objective (minimize cost) are expressed as equations and the linear programming routine finds the solution that satisfies the constraints and optimizes the value of the objective. The computed solution gives the amounts of each feed ingredient that should be used to obtain the least cost formulation for each particular feed.

Once the product has been shown to be safe for animal use and palatable to the animals, and process technology refined to minimize production cost, then the price to launch the product can be obtained from the formulators linear programming routine. The target market for the product launch should be the large, quality feed formulators - a large formulator will realize greater benefits from any price advantage of the SCP product. Providing this "price advantage" (at least initially) should be a marketing objective. This may result from the actual cost of the material on a percent protein basis or from cost savings resulting from a reduced requirement for expensive mineral and amino acid supplements in cases where these are included in sufficient amounts in the nutritional composition of the SCP product. These key ingredients include lysine, the sulphur containing amino acids, (methionine, and cystine), and B vitamins, biotin, riboflavin, folic acid and pantothenic acid.

Industry reluctance to use the product, in spite of successful feeding test results, will likely disappear as soon as one formulator uses the product if that formulator can gain an advantage over competitors in terms of materials cost and product quality. It is, therefore, very important to gain that first acceptance of the product by a formulator.

Advertising should emphasize the nutritional balance and high protein concentration which allow greater formulation flexibility, and that SCP products, genetically engineered for high protein, vitamin and mineral content raise the overall productivity and efficiency of animal production. While these two product dimensions are marketable, it is likely that the formulators will be most receptive to messages that stress the cost advantages of the product. However, the nutritional composition of the product will have a direct impact on the price that a formulator is willing to pay. A nutritionally superior product will have an advantage over even lower-priced protein sources in many instances. It is not imperative, therefore, that a SCP product be price competitive with other protein sources on a percent protein basis if there are nutritional advantages to using the single cell product from growth-promoting factors and vitamins, minerals, etc.

The best method to set a price at product launch time would be to include the nutritional data for a SCP product in the formulation linear program of a feed formulator, specify a price, and run the program to see if the program used the protein source of interest. A parametric analysis would indicate the amount of the protein source which would be included in the formulation at all different price levels. The optimal launch price is the highest price at which the protein product is still included in formulations in sufficient amounts to generate enough demand for the product.

AN EXAMPLE ECONOMIC ANALYSIS: SCP FROM WASTE SULPHITE LIQUOR

A process developed by the Czechoslovak Academy of Sciences in Prague, uses spent sulphite liquor from a pulp or paper mill as feedstock to a fermentation yielding a dried yeast product containing 45-50% protein. This yeast product has been tested in Czechoslovakia and found suitable for cattle feed at a concentration of 4-10% of the complete animal diet.

This process addresses a germane Canadian problem, disposal of pulp and paper mill waste liquors, thus an investigation of the economics of a plant in Canada was carried out. The proposed plant would have to be located close to a pulp mill, but the economic analysis was not specific to a single site. Rather a general economic/financial model was constructed which includes an allowance for site-specific costs such as materials, transportation, utilities, etc. The data used in this analysis was obtained through personal contacts and visits by the author. At the time the analysis was undertaken, the first commercial scale plant was still under construction although it is now in operation.

The process uses spent sulphite liquor containing 13-15% solids and 2-3% reducing sugars as the basic feedstock for the SCP plant. The assumed volume is 200,000 tonnes annually which is available from an unbleached pulp mill with an annual capacity of 200,000 tonnes of unbleached pulp (roughly equivalent to 2 million tons of wood). The feedstock has pH 1.8-2.5 and is at a temperature of approximately 80 °C. After stripping to remove sulphite and cooling to 35 °C, nutrients are added together with caustic potash and phosphoric acid, and the treated sulphite liquor is fermented with yeast (*Candida utilis*) in one of three parallel CHEPOS fermentors. The fermented medium is then separated, washed, evaporated and dried to produce a 92% dry matter solid containing 45-50% protein which is then packed in 25 kg bags for distribution. The separated liquor is returned to the pulpmill for disposal. The output rate of the plant is 25,000-28,000 tonnes of product annually.

The SCP product is a 92% dry matter meal of 45-50% protein that includes lysine, methionine, calcium, sodium, and phosphorus. The actual market value of such a product depends on detailed calculations of the impact of this new formula on the animal feed mixing problem. Various components of the SCP will have value in different types of feed--for example, lysine is an important component in poultry feed but less important in beef or dairy cattle feeds.

The component of the SCP product that is most easily valued is its protein content. Soyameal with a market price of $330/tonne (May, 1984) has a protein content of 48.4%, providing an estimate of the value of protein at $688/tonne. The SCP product therefore substitutes for protein in soyameal at a price of $310 - $344/tonne (45% - 50% protein). The additional ingredients may increase this price slightly, but at the same time a price discount may be necessary to gain entry with an unfamiliar product into a well established feed-mixing business.

The plant requires a site of approximately 0.8 hectares of serviced industrial land adjacent to a pulp or paper mill. It is anticipated that an Ontario municipality, attracted by the construction jobs plus the 42 permanent jobs that the plant offers, and a paper company, attracted by the opportunity to partially solve a difficult industrial waste problem will cooperate to provide a site. The availability of local incentives and grants suggests that the cost of this site is likely to be quite moderate and is assumed at $10,000.

Property taxes on the site plus the capital equipment investment, also provide further opportunities to benefit the local community. It is thought that a five year property tax waiver is a reasonable local development incentive, with a property tax rate of 2.5% of the market value of the plant plus site beginning in year six.

The construction cost of the Czechoslovakian plant was estimated at $21 million. An attempt was made to source the equipment and estimate the site preparation costs for Canadian construction leading to the conclusion that this $21 million may be a reasonable estimate of the extra cost of including the SCP complex in the liquid waste disposal facilities at a new mill. There is considerable uncertainty in this construction cost estimate, in particular little is known about the cost of the three CHEPOS fermenters required for the plant or of any royalty payments due to the owners of the design.

For the purpose of the analysis, the extra construction cost was assumed to be $21 million which is assumed to include equipment, installation, site preparation, utility hookups, transportation costs for equipment and licence fees or royalty payments. Plant construction is assumed to take 18 months with a 50% progress payment due in the first year and the balance in the second year.

Operating costs are summarized below:

Utility Cost Estimates in $/t Product are

Cooling Water	30.00	-	48.00
Process Water	5.25	-	7.50
Steam	9.60	-	12.00
Electric Power	28.70	-	37.31
Natural Gas	33.60	-	44.80
Total (minimum, maximum)	107.15	-	149.61

Total labor cost estimate:

Direct Labour $25.14 - 28.16/t product
Indirect Labour $125,970 p.a.

A maintenance allowance of 5% of the market value of the plant is assumed.

Total variable materials costs/t product:
$95.30 - 110.12

For a central Ontario location, transportation is estimated at 1,660,000/tonne miles annually. It is estimated that this can be contracted to a private trucker, or operated internally at a cost of $200,000 p.a.

Using the above data, a contribution analysis was carried out. (The contribution margin for a product is the excess of unit price over variable production cost. At a minimum, the contribution margin must be positive since this contribution must cover fixed (time dependent) costs as well as covering investment costs and generating a shareholder return). The unit price (or net revenue per tonne product) is assumed at $310 - 344/tonne and the variable cost/tonne range is found to be $227.59 - 287.89 (made up of utilities, labour, materials). The contribution margin/tonne at

$310/tonne is $22.11 - 82.41 and the contribution margin/tonne at $330/tonne is $42.11 - 102.41. For a scenario (production of 25,000 tonnes p.a. at $310/tonne), the contribution/tonne at $22.11 gives a total contribution of $552,750 and for a "best case" scenario (production of 28,000 tonnes p.a. at $330/tonne), the contribution/tonne at $102.41 gives a total contribution of $2,867,480.

From the total contribution must be subtracted indirect labour ($125,970), transportation ($200,000), and a monthly fixed utility cost that results from the block structure of utility rates. This fixed utility cost is estimated at $1,500/month or $18,000 p.a.

Net contribution is therefore in the range $208,780 to $2,523,510. This net contribution must also cover maintenance, property taxes, interest payments and profit. Assuming maintenance at 5% of capital investment and interest payments at 13% of capital investment, property tax at zero (initially) and zero profit, a contribution of $208,780 supports an investment of $1,159,889 while $2,523,510 allows an investment of $14,019,500 under the same conditions. Thus, if the plant plus land requires an investment of $14,019,500 at an interest rate of 13% with maintenance at 5% then at a net contribution of $2,523,510, the profit is zero. A "best case" scenario is summarized below.

Total revenue from 28,000 t at $330/t	$9,240,000
Less variable costs at $227.59/t	-6,372,520
Less annual fixed costs ($125,970+200,000+18,000)	-343,970
Less maintenance at 5% of $14,019,500	-700,975
Less interest at 13% of $14,019,500	-1,822,535
Leaves a profit of	0

The contribution margin analysis therefore reveals that if the plant can be built for less than $14 million, then there is a chance of a profitable operation under a "best case" scenario. If the needed capital investment exceeds $14 million, then a profitable operation will require some other means of support, either from government incentive payments or, perhaps more realistically, from waste disposal credits from the adjoining paper or pulp mill.

To investigate these other options in more detail, a financial/economic model was developed. A general financial/economic model of a commercial-scale SCP operation provides information on the economic feasibility of different types and sizes of SCP processes. The specific process must be developed and analyzed so that data are available as inputs for the general model. The model generates measures of the economic feasibility of the process. These could include return on investment (ROI), net present value (NPV) of after-tax cash flows, payback period, internal rate of return. The most acceptable of these as a measure of the financial desirability of a project is the net present value of after tax cash flows.

The difference between the simple contribution margin analysis performed earlier and the financial/economic model approach, is that the model looks at the process parameters as they evolve over time, while the simple contribution analysis examines only a single year. It is possible that an investment that is attractive over a twenty year period can make a loss in a single year.

The model was written using financial modelling software (Interactive Financial Planning System or IFPS, Execucom 1983) which facilitates rapid generation of results, easy manipulation of the data, and easy modification of the model and inputs to allow a thorough sensitivity analysis.

Because minor changes in the SCP production process can have pronounced effects on the costs involved, it is important that the process to be commercialized is defined as specifically as possible, and that scale-up effects from going from pilot plant to full-scale operation have been taken into account. As with any model, the outputs are only as reliable as the inputs.

For SCP from waste sulphite liquor, the contribution margin analysis (earlier) suggested that the Prague - SCP was borderline from an economic viewpoint. This analysis, therefore, begins with a "best case" scenario based on the following assumptions.

Yield: 28,000 tonnes p.a. SCP from 200,000 tonnes of sulphite liquor.

Percent protein in product: 50%

Substrate price per tonne: $0.

Waste disposal credit per tonne: $0.

Selling price of product: $688 per tonne protein

Tonnes of product sold per annum: 28,000

Chemicals cost per tonne of product: $95.30 (includes packing)

Direct utilities cost per tonne of product:$107.20

Direct labour cost per tonne of product: $25.14

Period costs: Indirect labour $125,970
Indirect utilities $18,000
Transportation $200,000
Maintenance 5% of market value of plant
Property tax 2.5% of market value of plane after 5 years

(Note that there is no allowance for physical depreciation of the plant - it is assumed that maintenance at 5% p.a. of market value can offset physical depreciation).

Tax rate: Ontario tax structure.

Capital cost allowance rate for equipment: 0.2

Additional taxes and credits: None.

Expected financing: 100% loan for land, captial equipment and working capital at 13% fixed rate repayable over 20 years.

Inflation: Inflation rate 6% in perpetuity. All costs and prices inflate at this rate.

Cost of Capital: Assumed 9% for present value calculations.

Capital investment: Assumed at construction cost of plant plus land inflating at inflation rate, plus changes to working capital. The capital investment is, therefore, an approximation of the market value of the SCP operation assuming that maintenance charges are sufficient to offset physical depreciation.

Working capital to finance operations: Assumed at the market value of some month's production SCP.

Plant Startup: It is assumed that the plant is built in 18 months. First year production is zero, second year production is assumed at 14,000 tons. The land cost plus half the construction cost is borrowed in the first year, the balance of the loan in the second year. Period costs are zero in the first year and one half of annual rates in the second year.

The financial/economic model was first run to show "best case" economic analysis for a twenty-year horizon. The results from the run are quite discouraging. The after tax cash flow (ATCF) is negative through year 11, and the net present value of after tax cash flows is negative $24.4 million over twenty years (recall that this is a "best case" analysis).

In considering how this situation might be improved, one line of attack is through some form of waste disposal credit. Existing paper or pulp mills already have elaborate waste treatment facilities, but if the SCP plant were attached to a new mill, then the investment in the SCP plant would offset otherwise needed investment in waste treatment facilities. Under these conditions, it would be profitable to the mill owner to provide a credit to the SCP plant for waste disposal. A waste disposal credit of about $10 per tonne of sulphite liquor feedstock is sufficient to generate an attractive economic situation. In this case, after tax cash flow is now positive in the first full year of production and the twenty-year horizon yields a net present value (of ATCF) of - $3.6 million and annual rates of return peak at over 12% in year 11.

At $10 per tonne sulphite liquor, the pulp or paper mill is spending $2 million ($10 x 200,000 tonnes) annually on the SCP plant. This level of subsidy from any other source would have a similar economic impact and an amount slightly above this is needed to make the SCP plant attractive.

The simple contribution analysis suggests that the fundamental problem is that the variable costs are too high and hence the contribution uneconomically low. It seems unlikely that utilities or chemicals costs can be reduced very much, but it may be possible to reduce labour costs. If direct labour can be reduced by 50%, then the situation is improved slightly. One further possible economy is to ship the product in bulk (FOB plant), avoiding the packaging cost. Considering this on top of the labour cost reduction and the waste disposal credit of $10/tonne leads to an investment that looks quite attractive. The return on investment reaches a maximum of close to 16% and the twenty-year net present value is $4.5 million. If the waste disposal credit is cut to $5/tonne, however, the economics changes markedly and the plant looks unattractive.

A second approach to improving the economic picture is to try to reduce the capital investment required. To investigate the impact of this, a reduction of the captial cost by 50% is examined and the results yield a twenty-year present value of after tax cash flows of $4.8 million indicating that the cash flow can support the reduced level of investment. At an investment level of $15 million, however, the twenty-year present value is negative $7.54 million.

CONCLUSIONS

It does not appear that the Prague, SCP from waste sulphite liquors process can be built and operated economically in Canada at this time, unless either, the cost of the plant can be heavily subsidized by grants or equipment allowances, or the plant can be subsidized by a paper or pulp mill operator. The needed capital grant is of the order of $8-10 million, while the needed operating subsidy is of the order of $2 million p.a. under the most favourable circumstances. In fact, levels of subsidy much higher than this are likely to be required for profitable operation of the SCP plant. The actual profitability achieved will depend on the achieved output level of SCP, the percent protein in the SCP product and its quality,. and actual cost realizations.

In many other countries, subsidies are readily obtainable from government. This is especially true in countries where self-sufficiency in food production is an important national objective (e.g. the USSR and several eastern European countries). Canada, however, is a major exporter of grains and, given the general glut and low price levels in the world grain markets, there is little incentive for the Canadian government to subsidize production of SCP animal feeds that substitute for grains. Further, the current depressed state of Canadian agriculture and the farm bankruptcies that continue to fill our newspapers suggest that sizable government subsidies to SCP producers would be a political liability.

However, it is important to keep in mind that pulp and paper mill waste liquors cannot be discarded but must be treated for pollution control. This treatment is usually a biological conversion of sugars from the liquor to a sludge which is a nuisance that has to be disposed of. It would, therefore, make sense to convert the sugars from the liquor to a feed supplement which has a nutritional and commercial value. This paper shows that this process becomes economical if a credit for the cost of water treatment is included.

In summary, it appears that a large scale plant for production of SCP from pulp and paper mill effluents would be attractive in Canada, even though Canada does not have a big need for alternate animal feeds. In the meantime, large scale SCP production facilities are being built elsewhere in countries where other, non-economic, factors make it advantageous to produce their own animal feed supplement. Often this is motivated by a need to reduce imports to conserve hard currency.

REFERENCES

Ericsson, M., Ebbinghaus, L., and Lindbloom, M., "Single-Cell Protein from Methanol: Economic Aspects of the Norprotein Process", J. Chem. Tech. Biotechnol, 31 (1981), 33-43.

Interactive Financial Planning System, Execucom, Atlanta, Georgia (1983).

Kosaric, N., Bell, P.C., Cosentino, G., Magee, R., Turcotte, G. and Purcell, A., "Non-Conventional Microbial Food: Industrial and Economic Possibilities for Canada", Technological Innovation Studies Program Research Report 96, Office of Industrial Innovation, Department of Regional Industrial Expansion, Government of Canada (July, 1984).

SEXUAL CYCLE INDUCTION IN THE WHITE ROT FUNGUS *SPOROTRICHUM PULVERULENTUM*

Manuel Raices and O. Rolando Contreras
Microbiology Laboratory
Industrial Microbiology Department
CENIC
P.O. Box 6990
City of Havana
Cuba

SUMMARY

Induction of fruiting bodies, including basidia and basidiospores, was observed after 5 days incubation in one strain of the white rot fungus *Phanerochaete chrysosporium* (formerly called *Sporotrichum pulverulentum*) using an agar medium containing low nitrogen levels, walseth cellulose as the sole carbon source, saponin to restrict colony size, and mineral salts.

INTRODUCTION

The white rot fungus, *Sporotrichum pulverulentum* Novobranoca sp., has been extensively studied for its ability to degrade cellulose and lignin (Eriksson, 1983). *Phanerochaete chrysosporium* Burds ME-446 (ATCC 34541) and *Sporotrichum pulverulentum* Novobranoca Sp. (Atcc 32629) have been shown to be closely related although not entirely identical, on the basis of DNA hybridization (McDonald et al., 1983).

Genetic studies of these strains allow the isolation of new types through meiotic recombination and gene segregation (Gold, 1979). For this reason, induction of fruiting body and basidiospore production in this species is desirable. Studies of the sexual cycle in *Sporotrichum pulverulentum* show that it is under the control of nitrogen levels and the carbon source used. The effects of low nitrogen levels, in conjunction with a carbon source like walseth cellulose, on rapid fruiting body formation of *Sporotrichum pulverulentum* are reported in this paper.

MATERIAL AND METHODS

Organism: *Sporotrichum pulverulentum* Novobranova (ATCC 32629) was kindly provided by Dr. K.E. Eriksson, of Swedish Forest Product Research Laboratory, Stockholm, Sweden.

Culture Media: Malt extract agar contained malt extract 20 g/L and agar 20 g/L. The nutrient salt medium (Eriksson et al. (1980)) had the following composition per liter: KH_2PO_4, 0.6 g; $MgSO_4 \cdot 7H_2O$, 0.5 g; K_2HPO_4, 0.4 g; $CaCl_2$, 0.048 g; ferric citrate, 0.012 g; $ZnSO_4 \cdot 7H_2O$, 0.0066 g; $MnSO_4 \cdot 4H_2O$, 0.005 g; $CaCl_2$, 0.001 g; $CuSo_4 \cdot 5H_2O$, 0.001 g; thiamine, 0.0001 g; the nitrogen sources consisted of $NH_4H_2PO_4$ and urea in concentrations of 2 g and 0.86 g respectively for the high nitrogen medium, 0.123 g and 0.004 g for the low nitrogen medium. The carbon source consisted of walseth cellulose (Walseth, 1952) at 1%. Saponin (1 g/L) was added to restrict colony size. pH was adjusted to 5 with 0.1 M H_2SO_4

Conidiospore Production: Malt extract slants were inoculated and incubated at 37 °C for 5 days. Conidiospores were suspended in sterile saline.

Fruiting Body Production: A double layer method was used in which the bottom layer (20 ml/plate) contained 1.5% agar (Difco) nutrient salt and 0.1% saponin. 30 conidiospores were plated on this medium and after one hour the upper layer (10 ml/plate), consisting of 0.7% agar (Difco) with nutrient salt (high or low nitrogen as indicated) and 0.1% saponin was added (Raices, 1982). Plates were incubated inverted at 30 °C. Macroscopic and microscopic observations were made daily.

RESULTS AND DISCUSSION

After five days incubation, some pale gray spots on the petri-dish cover were observed, opposite colonies, resembling these colonies in shape and showing the same surface globular structures (Figure 1). These dish-cover spots were washed with 2 ml of sterile water, and these suspensions were collected and used for microscopic observation. Many kidney shape spores with low refringence were observed, unlike conidiospores normally found in *Sporotrichum pulverulentum* cultures.

Microscopic preparations from globular structures of these same colonies, showed the existence of mycelial zones composed of basidia, probasidia and high concentrations of kidney shaped spores, some of them connected with each of the four basidial extremes. These results led us to identify the kidney shaped spores as basidiospores, indicating the induction of a sexual cycle in this species (Figures 2a-2b).

Not all the colonies developed fruiting bodies in these culture media. These colonies showed high variability with respect to size and shape, but the texture and color of all of them were the same. The existence of few whole structures and the accumulation of basidiospores on the petri-dish covers indicated a low stability for the structures connecting basidiospores to the basidia.

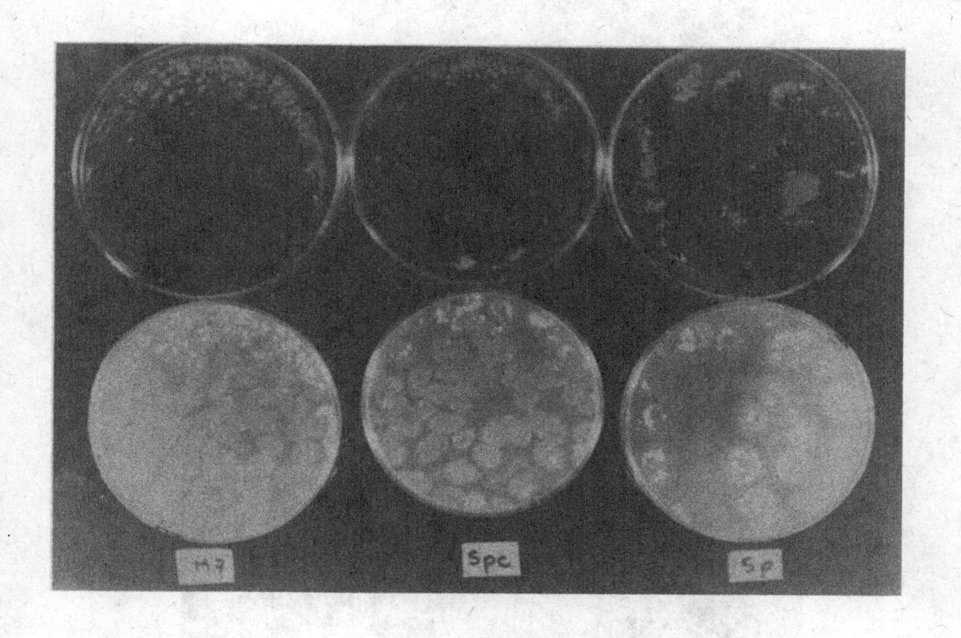

FIGURE 1: Petri-dish after five days incubation

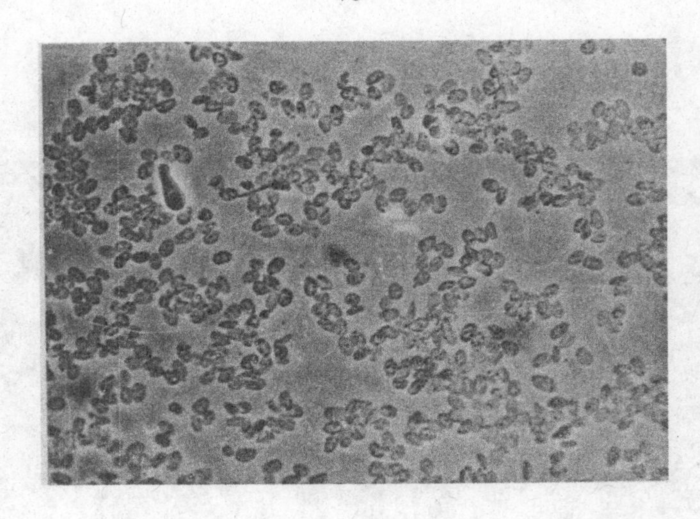

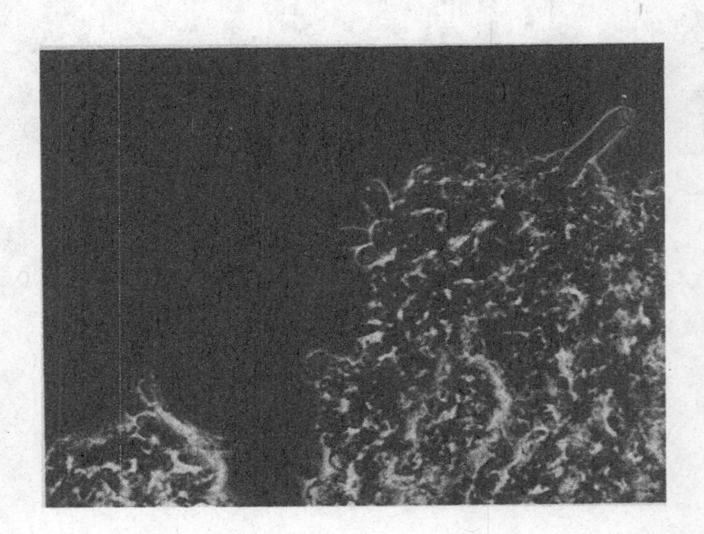

FIGURES 2a & 2b: Basidiospores evidencing the sexual cycle induction in *Sporotrichum pulverulentum Novobranova* (ATCC 32629)

Table 1 shows the number of colonies and fruiting body formation was higher in the low nitrogen medium, suggesting that fruiting is under nitrogen metabolite repression as reported by Gold (1979).

TABLE 1: Nitrogen influence on the fruiting body induction

Nitrogen level in salt medium	Total number of colonies (30 col/plate)	Total number of fruiting bodies
High	295	2
Low	312	129

The fungal strain developed from one basidiospore has been purified by micromanipulation in order to study the life cycle of this species. Further studies concerning these genetic features are being carried out.

ACKNOWLEDGEMENTS

We express our sincere thanks to Dr. El Odier, Laboratoire de Microbiologie, Institute National Agronomique, Paris Grignon, France, for suggestions and critical reading of the manuscript. This work was supported by IFS Grant No. E/734-1.

REFERENCES

1) Eriksson, K.E. (1980). Development of biotechnology within the Pulp and Paper Industry, 27th International Congress, I.U.P.A.C.

2) Eriksson, K.E., Johsrud, S.C., Vallanders, L. (1983). Arch. Microbiol. 135: 161-168.

3) Seligy, V.L. (1984). Biotechnology in the Pulp and Paper Industry London 12th, 13th, 14th, September.

4) Gold, M.H., Cheng, T.M. (1979). Arch. Microbiol. 121: 37-41

5) Eriksson, K.E., Grunewald, A., Vallanders, L. (1980). Biotech. and Bioeng. Vol. 22, 363-376.

6) Walseth, C.S. (1952). Tappi 35: 228-233.

7) Raices, M.R. (1982). Tesis de Diploma. Facultad de Biologia, Universidad de La Habana.

ENZYME ACTIVITIES DURING *PLEUROTUS OSTREATUS* GROWTH ON WHEAT STRAW UNDER CONTROLLED CONDITIONS

G. Giovannozzi Sermanni, M. Luna, M. Felici, F. Artemi and M. Badiani

Agric. Chem. Inst.
University of Tuscia
Viterbo, Italy

INTRODUCTION

Biomass generation of lignocellulosic materials has an increasing economic importance for all countries, where agricultural residues can constitute valuable sources of organic substrates for organisms, widely present in nature, which are capable of degrading cellulose and lignin.

A solid state fermentation has many advantages over stirred or liquid fermentations: simple preparation of the medium, simple equipment, efficient oxygen and carbon dioxide exchanges if the substrate density is suitable, and low cost of the drying process. On the other hand, heat dissipation and control of the fermentation parameters, such as pH, can be more difficult (1,2,3).

Chemical delignification improves the biomass fermentation because it improves accessibility to the cellulose, but it adds another step to the process, thus increasing the cost. Bioconversion by organisms having ligninolytic activities could avoid the need for chemical treatments, and is, therefore, of some interest.

An understanding of the biochemistry of ligninolytic organisms is important for achieving the greatest efficiency in their use, but in spite of the importance of this subject, research reports on it are scarce. For this reason, we have studied *Pleurotus ostreatus,* a fungus showing this capability, and the effects of CO_2 on its growth, since *Pleurotus ostreatus* has been reported to be resistant to CO_2 concentrations up to 20 to 25% (4).

The data shown here concern several aspects of the relationships between carbon dioxide level, and cellulose and lignin degradation, protein production and the activities of some enzymes, as summarized in Figure 1. The main macromolecules, lignin and cellulose, are attacked by hydrolytic (cellulase for cellulose) and oxidative (polyphenoloxidases for lignin) reactions.

Oxygen is utilized to produce carbon dioxide and water by means of respiration, but it can also be utilized in the presence of unfavourable environmental conditions or senescence. In fact it has been shown that the formation of a toxic species of oxygen in plant cells, the superoxide anion, is very dangerous for the integrity of membrane systems, and that it can be destroyed by superoxide dismutase (SOD) (5). If it is postulated that such behaviour also occurs in mycelia - in fact superoxide anions are produced by *Coriolus versicolor* (6) - the SOD level in the mycelium could be an indication of the physiological status of its mycelia. To complete the scheme, it is possible that carbon dioxide, arising from respiration, could be utilized for carboxylating reactions (Figure 1).

Therefore, we tested cellulase, laccase, isocitric dehydrogenase (ICDH), phosphoenol pruvate carboxylase (PEPCase), superoxide dismutase (SOD), total N content, protein and ash.

MATERIALS AND METHODS

Substrate

Four kg of wheat straw were shredded into 4-cm pieces and washed several times with tap water until clear. The shredded straw was squeezed and then dipped in a tank containing the following solution: sucrose 0.1%; $(NH_4)_2HPO_4$, 0.1%; $MgSO_4$, 0.05%; CH_3COOH, 0.1%; ZnCl, 2 ppm; $CuSO_4$, 2 ppm. After several hours of soaking, excess liquid was drained off, and the solid substrate (72% water, pH = 6.0) was sterilized in heat-resistant plastic bags at 120°C, for 30 minutes. After cooling, it was inoculated, under a sterile hood, with 3% (w/w) of mycelium of a commercial strain of *Pleurotus ostreatus,* put in a tray, and placed in the fermentor.

Fermentation Equipment

The home-made fermentor (80 cm x 90 cm x 95 cm) was built with perspex sheets (Figure 2). Mobile walls (A), in the inside, supported up to 3 plastic trays (B) with bored edges and bottoms. Two baffles (C) allowed uniform air changes from a pipe filled with sterile-wool, (D) at the top of the fermentor, to a pipe at the bottom connected to a suction pump (E). Rubber gloves (F) mounted on the front panel allowed sterile handling of the inner material without altering the growth conditions. A box, supplied with two hermetic seals (G), allowed samples to be removed or treatments to be carried out, without disturbing the conditions inside the fermentor. Electronic thermometers continuously recorded the temperatures of the fermenting material and of the air at the different levels. Sterilized rubber tubes (H) permitted the gaseous phase from the inside of the fermenting substrate to be aspirated and the levels of carbon dioxide, or other volatile compounds, to be determined.

Before its use, the fermentor was sterilized by means of formalin vapours for 24 hours; then formalin was replaced with sterile air until the odour disappeared. The temperature of the fermentor was controlled by an electric blanket equipped with a vertex thermometer.

After the tray with the inoculated substrate was placed in the fermentor, the front panel (made airtight by soft rubber strips) was closed and the mycelium was allowed to grow, without changing the air in the fermentor, while CO_2 evolution and related biochemical activities were followed. In these conditions, during 15 days, the pH remained roughly constant and no contamination occurred.

CO_2 and Ethylene Determinations

CO_2 and ethylene concentrations were determined by using specific reacting tubes (Drager, A.G. Lubeck, Germany) through which air was pumped; CO_2 concentration was determined by reference to the colour reaction of carbon dioxide with a hydrazine compound, the consumption of which was indicated by a change in colour of a redox indicator (crystal violet). Ethylene concentration was determined by the reaction of ethylene with a molybdate/palladium reagent.

Sample Preparation

Ten grams of each sample were stirred with 50 mM sodium phosphate buffer, pH 7.0, in an Ultra turrax stirrer for 2 minutes, filtered through gauze, centrifuged at 30,000 G for 30 minutes, and the supernatant used as source of enzymes. All procedures were carried out at 4 °C.

Enzyme Assays

SOD activity was assayed according to Dhindsa et al. (7) by following the inhibition of the UV-catalyzed reduction of nitro-blue tetrazolium by means of riboflavin at 560 nm.

PEPCase was assayed following Di Marco et al. (8) and the activity was calculated according to Solomonson (9). Peroxidase activity was assayed according to Valenta et al. (10) which involves the reaction between guaiacol and H_2O_2 at 470 nm.

ICDH activity was assayed by measuring absorption changes at 340 nm due to the reduction of NADP as it reacted with L-isocitrate to form α-ketoglutarate (11).

Cellulase was assayed according to Goodenough and Man (12) by determining the reducing sugars formed, with the Nelson method, after incubation with carboxymethyl cellulose.

Laccase was assayed using a simple and rapid method, involving guaiacol oxidation at 465 nm (13), instead of the accurate HPLC method (14) less suitable for numerous assays.

Protein was assayed by the BIO RAD dye reagent method.

Total nitrogen was determined by the Kjeldhal method.

RESULTS AND DISCUSSION

Similar changes in carbon dioxide concentrations occurred in the fermenting solid material at two different temperatures (Figure 3). In agreement with previously published results (4) the maximum CO_2 concentrations were reached in 6-7 days regardless of the absolute CO_2 levels reached. The temperature curve at 20 °C was similar to the curves for CO_2 (Figure 4) and agreed with temperature curves obtained for *Pleurotus ostreatus* growing in plastic bags where the temperature reached a maximum 5 days after inoculation (15). The inversion of the carbon dioxide curves at the two different temperatures does not appear to be due to CO_2 levels since the maximum concentrations, recorded at the 5th-6th day, were clearly different (5% and 10%). Anoxia conditions or induced senescence could be responsible for this behaviour. In fact 0.4 ppm of ethylene were found, 8-9 days after inoculation, in the inner atmosphere of the substrate, where the highest CO_2 concentration and the greatest temperature variations were observed. Figure 5 shows the nitrogen content of the microbial biomass plus substrate during mycelial growth. The content is expressed in relation to the ash content of the substrate so as to avoid an error due to the decrease in total organic matter resulting from respiration. The concentration of nitrogen increased until the 6th day after inoculation and became constant thereafter. The data indicate an increase in the total nitrogen content, which agrees with published data suggesting that *Pleurotus ostreatus* may be able to fix atmospheric nitrogen (16, 17). This result is provocative because eucaryotic organisms, such as *Pleurotus ostreatus* are said to be unable to fix nitrogen. The respiration rate seems to be strictly related to the nitrogen increase because an excellent linear relationship was obtained between nitrogen increment and CO_2 evolution (Figure 6). A vigorous growth of Pleurotus in liquid nitrogen-free medium has been observed (A. Ponente and I. Cacciari, personal communication).

The steady-state level of the nitrogen content after the 6th day suggests that nitrogen-fixing capability was strongly inhibited by that time and may indicate the presence of some limiting factor.

The cellulase activity was maximum at the 6th day and declined slowly afterwards, while the reducing sugar levels did not vary greatly except for reaching a peak at the 11th day after the inoculation (Figure 7). The same behaviour has been reported for invertase production, which ceases after 5 days, with *Aspergillus awamori* grown in a stirred vesssel (18).

The ratio cellulase/reducing sugars showed a constant increase until the 11th day except a peak at the 5th day. Since it is known that cellulase production is regulated by cellobiose and glucose (3), the constant ratio could be due to the regulation itself.

Laccase, which is one of the main enzymes involved in the biochemistry of lignin (19, 20, 21), exhibited its maximum activity at the same time as did cellulase activity (Figure 8), so that these two groups of enzymes had similar activities during the fermentation. Therefore, cellulose and lignin degradations appeared to be related, in confirmation of our previous results (22,23).

Unexpectedly, the ICDH activity increased continuously during mycelial growth (Figure 9). This increase did not correlate with CO_2 evolution and could indicate that the decrease in the rate of CO_2 evolution may be due to physicochemical (solubilization, etc.) or biochemical (carboxylations) effects.

As previously stated, the laccases oxidize aromatic compounds to phenoxyradicals and water in the presence of oxygen, while spontaneous radical reactions continue to occur with aromatic groups in macromolecules (24). It has been shown that lignin degradation requires oxygen and the amount of oxygen available to the culture greatly affects the rate of the biodegradation of the lignin (25, 26). Therefore, by decreasing the oxygen tension, the rates of respiration and the lignin degradation could be decreased and stress conditions may appear.

During the fermentation under our experimental conditions, the SOD activity increased after one week of growth (Figure 8), reaching its maximum at the 11th day. The presence of a peroxide-requiring extracellular enzyme able to cleave C-C bonds has been described in another mycelium-forming organism (27, 28). SOD could be involved in the production of H_2O_2, which in turn might be utilized for a lignin degradation process in the absence of laccase activity, while the concentration of superoxide anions decreases. The increase of SOD activity immediately after the drop in laccase activity could be compatible with the above model, but until now our analyses of peroxidase activities do not support such an explanation.

The irregular PEPCase activity found by us in the fermenting substrate does not explain entirely the decrease in CO_2 concentration in the biomass and the presence of other carboxylating reactions may be a reasonable hypothesis.

The laccases are very interesting in their biochemical and biotechnological properties (29, 30). Therefore, in another research project, having as its aim amplification of laccase expression in a prokaryote, *Pleurotus ostreatus* has been grown in submerged culture and the nucleic acids have been purified from mycelium homogenates. After separation of RNA from DNA, the RNA fraction was enriched in mRNA by affinity chromatography on oligodT-cellulose. The ability of the mRNA fractions to direct protein synthesis *in vitro* has been tested both in rabbit reticulocytes and in wheat germ lysate systems in the presence of ^{35}S methionine. The translation products have been analyzed by SDS-polyacrylammide gel electrophoresis and immunoreactivity against antilaccase antibodies. The synthesized proteins had molecular weights between 50,000-100,000 daltons (V. Buonocore, personal communication).

CONCLUSIONS

The data presented indicate that:

a) the home-made fermentor described here is a useful device for studying the transformations of solid materials (in controlled conditions) and for clarifying its physiological and biochemical aspects of mycelial growth with a minimum disturbance of the geometry of the fermenting mass and of the inner environment.

b) 5-6 days after inoculation the CO_2 concentration, cellulase and laccase activities declined, the SOD increased and the total nitrogen content reached a steady-state.

c) the nitrogen-fixing activity appeared to be dependent on respiration.

d) the maximum CO_2 concentration produced by *Pleurotus ostreatus* seemed to be dependent on the fermentation conditions and on the strain.

Research on the biochemistry of ligninolytic organisms is of value for indicating the best organisms and the best fermentation conditions for efficient bioconversion.

REFERENCES

1) Laukevics J.J., Aspite A.F. Viesturs U.E. and Tengerdy R.P.; (1984) Biotech. Bioeng. 26: 1465-1474

2) Hesseltine C.W.; (1977) Proc. Bioch. July August.

3) Cannel T. and Moo Young M.; (1980) Process Biochem. August September 24-28.

4) Zadrazil F.; (1975) Eur. J. Appl. Microbiol. 1: 327-335

5) Elstener E.F.; (1982) Ann.Rev.Plant Physiol. 33: 73-96

6) Amer G.I. and Drew W.; (1981) Dev. Ind. Microbiol. 22: 479-484

7) Dhindsa R.S., Plumb-Dhindsa P. and Thorpe T.A. (1981) J.Exp.Bot. 32, 126: 93-101

8) Di Marco G., Grego S., Pietrasanti T., and Tricoli D. (1976) J.Exp.Bot. 27: 725-734

9) Solomonson L.P.; (1975) Plant Physiol. 56: 853-855

10) Valenta L.J. et al.; (1973) J.Clin. Endocrin.Metab. 37,4: 560

11) Bergmeyer H.U.; Methods of Enzymatic Analysis 2nd ed. Vol.1: 264-267 Academic Press Inc.

12) Goodenough P.W. and Maw G.A.; (1975) Physiol. Plant Pathol. 6: 145-157

13) Luna M., Poerio E., and Badiani M.; (1983) Agrochimica XXVII, 1: 44-50

14) Badiani M., Felici M., Luna M., and Artemi F.; (1983) Anal.Biochem. 113: 275-276

15) Laborde J.; (1981) Bull. FNSACC 9: 3-249

16) Ginterova A., Gallon J.R.; (1979) Bioch. Soc. Trans. 7,6: 1293-1294

17) Rangaswami G., Kandaswami T.K., Ramasamy K.; (1975) Current Science 44, 11: 403-404

18) Silman R.W.; (1980) Biotech. Bioeng. 22: 411-420

19) Kuroda H., Shimada M., and Higuchi T.; (1975) Phytochemistry 14: 1759

20) Thoms A., and Wood J.M.; (1970) Biochemistry 9: 337

21) Crawford R.L., Kirk T.K., Harking J.M., and McCoy E.; (1973) Appl. Microbiol. 25: 322

22) Giovannozzi Sermanni G., Basile G., and Luna M.; (1978) 10th Int. Cong. on the Sci. and Cultiv. of Edible Fungi: 37-53 Bordeaux

23) Giovannozzi Sermanni G., and Luna M.; (1981) Mus.Sci. XI, 2nd part: 485-496 Sidney

24) Felici M., Artemi F., Badiani M., and Luna M.; (1983) III Europ. Symp. Org. Chem. 5-9 Sept. Canterbury

25) Bar-Lev S.S. and Kirk T.K.; (1981) Bioch. Biophys. Res. Comm. 99: 373-378

26) Leisola M., Ulmer D., and Fiechter A.; (1983) Eur. J. Appl. Microbiol. Biotechnol. 17: 113-116

27) Faison B.D., and Kirk T.K.; (1983) Appl. Environ. Microbiol. 46: 1140-1145

28) Tien M., and Kirk T.K.; (1983) Science 221: 661-663

29) Felici M., Artemi F., and Luna M.; (1985) J. of Chrom. 320: 435-439

30) Felici M., Luna M., Artemi F., and Badiani M.; (1984) Agrochimica 28: 251-256

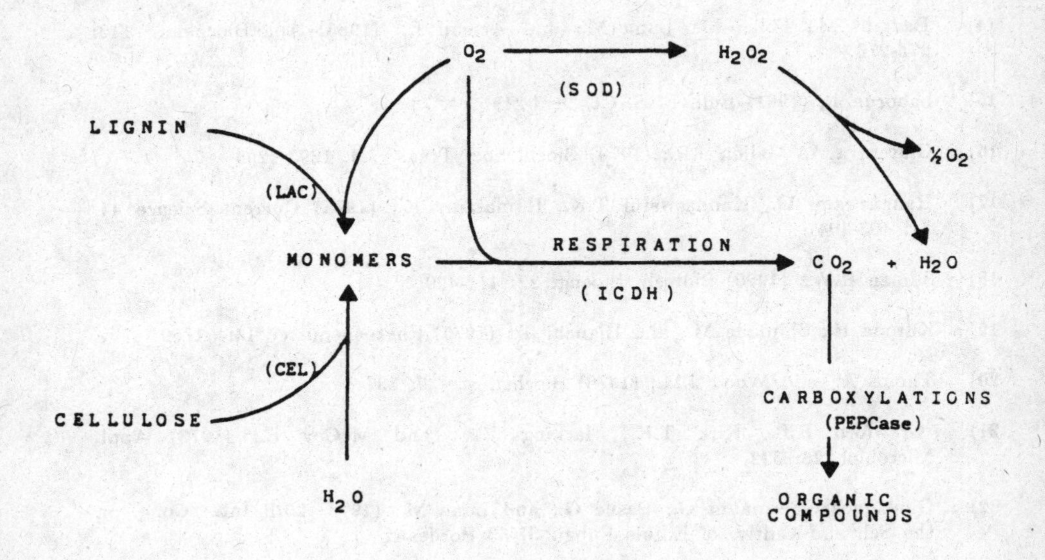

FIGURE 1: Relationships between some compounds and enzymatic reactions concerned with lignin and cellulose degradation.

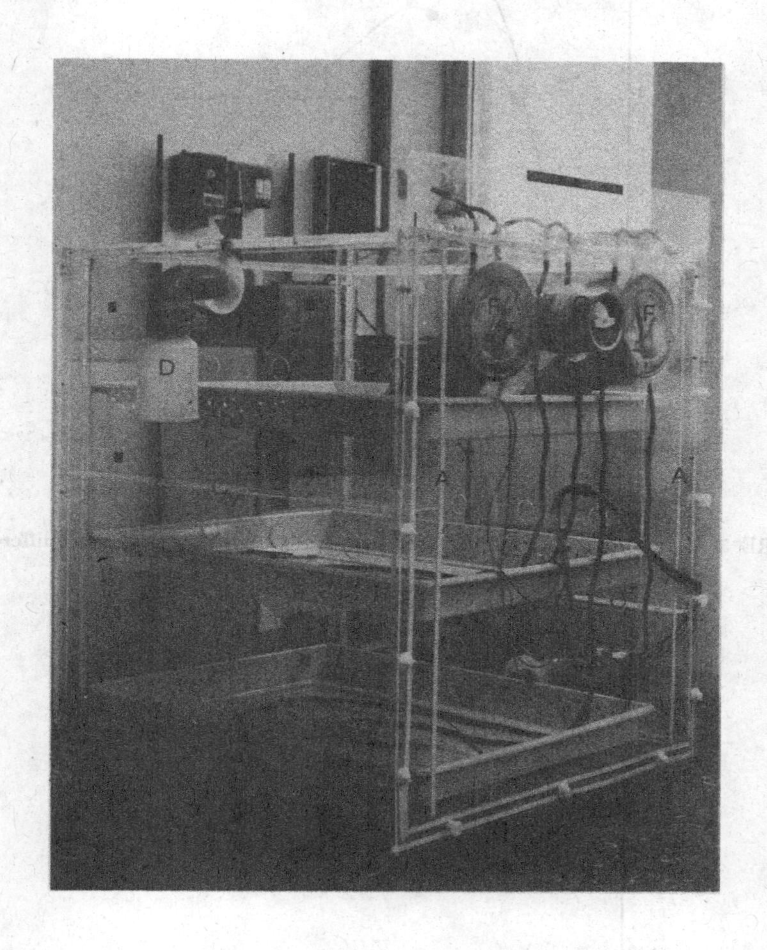

FIGURE 2: The fermentor

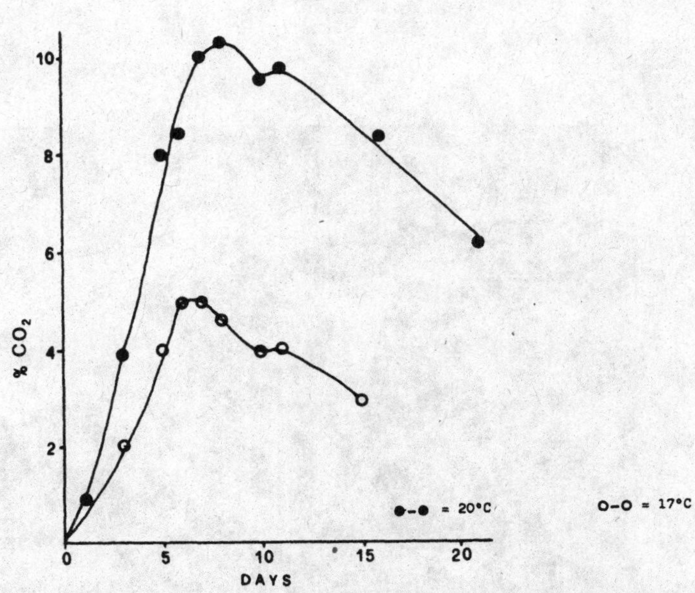

FIGURE 3: CO_2 concentration in the fermenting substrate at two different air temperatures

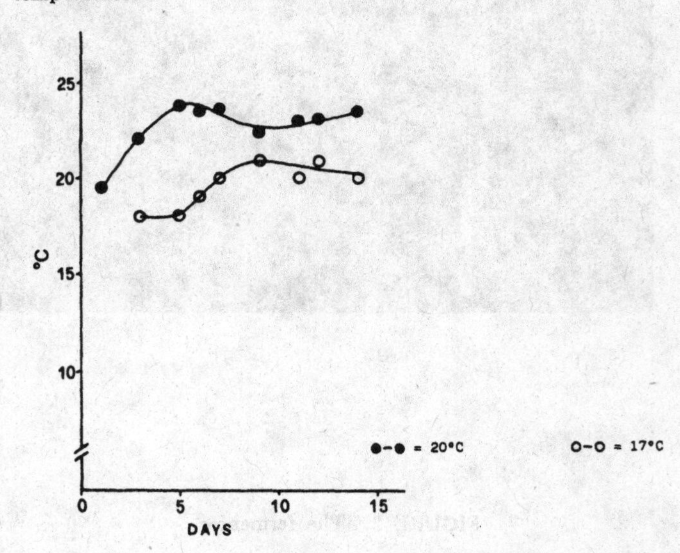

FIGURE 4: Temperature of the inside of the fermenting substrate at two air temperatures

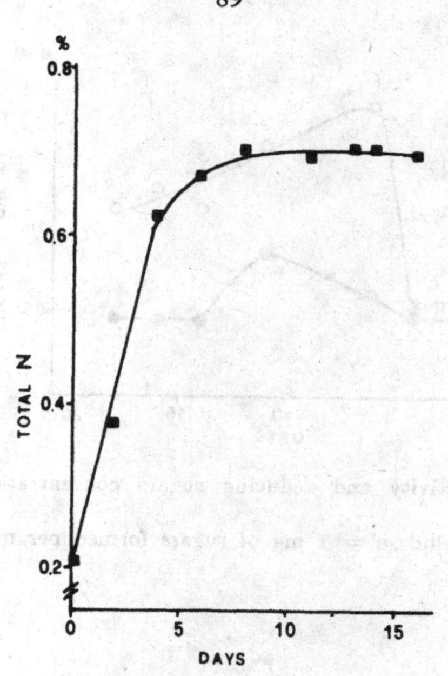

FIGURE 5: Percentage of total N calculated in relation to its ash content of the substrate

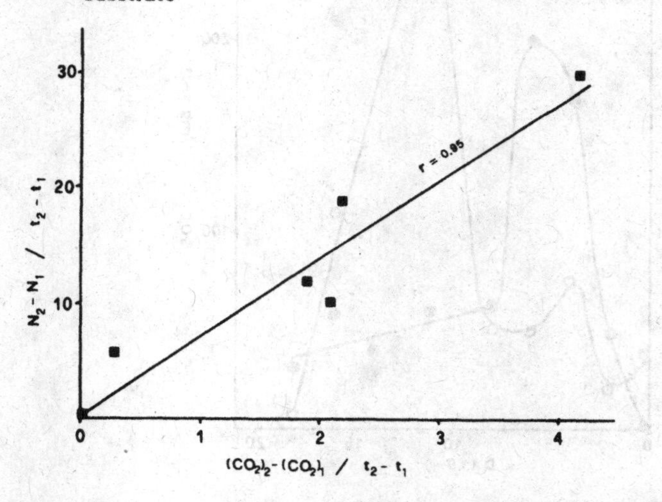

FIGURE 6: Nitrogen increment plotted against incremental CO_2 evolution

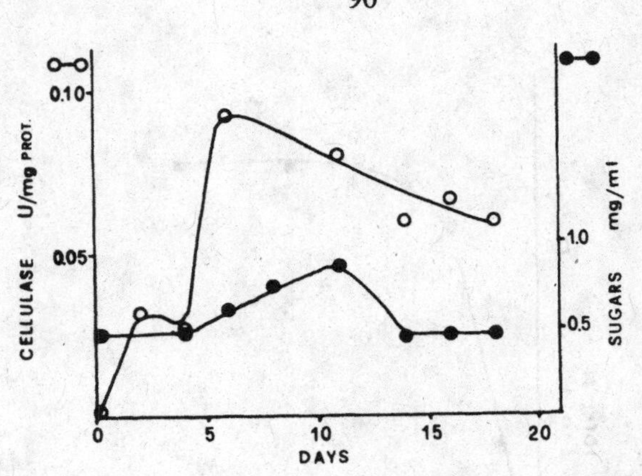

FIGURE 7: Cellulase activity and reducing sugars concentration in the fermenting substrate
1 Unit of cellulase = 1 mg of sugars formed per min. of incubation

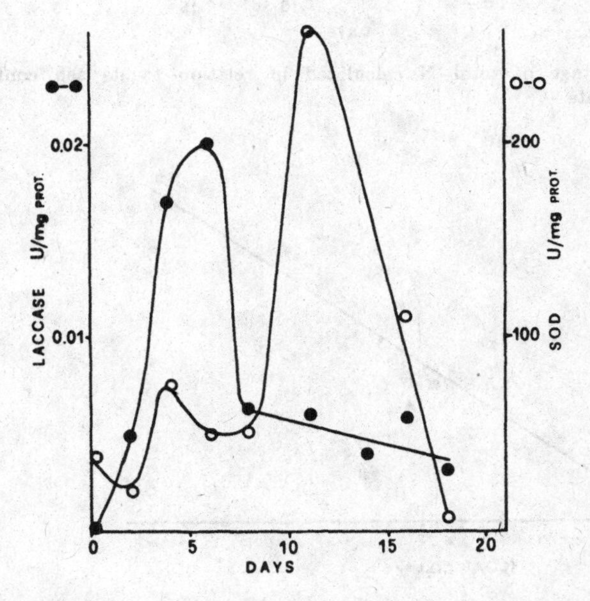

FIGURE 8: Laccase and superoxide-dismutase activity in the fermenting substrate
1 Unit of laccase = 1 E at 465 nm per min.
1 Unit of cellulase = 0.5% of inhibition of the reaction rate of the assay

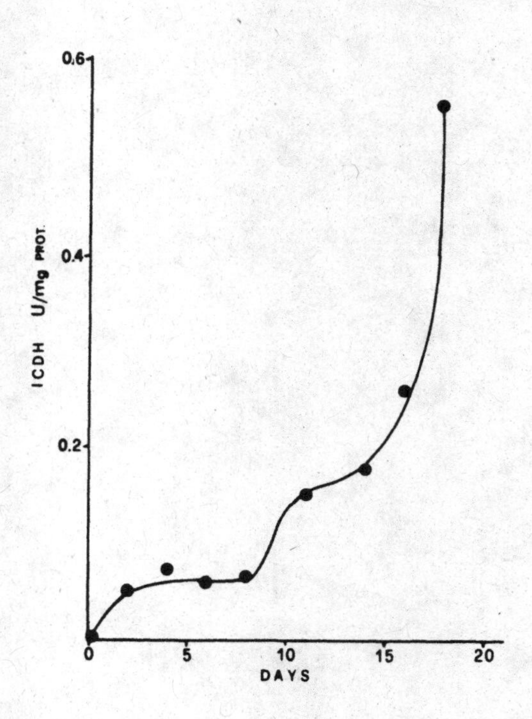

FIGURE 9: Isocitrate-dehydrogenase activity in the fermenting substrate
1 Unit of ICDH = 1 micromole of NADPH formed per min.

REGULATORY ASPECTS OF MICROBIAL BIOMASS PROTEINS AS FEED INGREDIENTS

S.K. Ho
Feed and Fertilizer Division
Canada Department of Agriculture
Ottawa, Ontario, Canada
K1A 0C6

INTRODUCTION

In Canada, commercial feeds and feed ingredients are regulated under the Feeds Act and Regulations. Any new feed ingredients, including novel protein sources derived from microbial biomass, must be proven safe and efficacious for their intended purpose before being offered for sale in Canada. The Feed and Fertilizer Division of the Canada Department of Agriculture is charged with the responsibilities of administering this legislation.

LEGISLATIVE FRAMEWORK

A new set of Feeds Regulations came into effect in 1983. Schedule IV of these Regulations lists the standards of composition and quality of all single ingredient feeds permitted to be used in Canada. Microbial biomass proteins, if approved, will be listed in Part II of the Schedule, and will require registration before being offered for sale. The Feeds Regulations have safety standards (Appendix I) in place, as well as guidelines of what constitutes satisfactory evidence (Appendix II) in support of a petition for the defining and listing of a new ingredient in Schedule IV.

EVALUATION GUIDELINES

The regulations stipulated in Appendix I and II constitute the legislative framework. But, as a feed regulatory official, one must exercise professional discretion in the interpretation of these Regulations when one is dealing with the acceptability of novel ingredients such as microbial biomass protein, and especially in deciding the extent of evidence that will be considered "satisfactory". This is the reason why, up to this point in time, the traditional route of setting up firm evaluation guidelines has not been followed. Instead, flexibility is built into the mechanics of the system, and submissions will be handled on a case-to-case basis.

From what is known so far, it appears that some parameters such as the ones outlined below could influence the interpretation of the Regulations:

- substrate from which the microbial biomass protein is derived;
- microbe being used;
- type of process used; and
- intended end use of the microbial biomass protein.

These parameters will guide the regulatory official, as well as the manufacturer, in determining what questions should be asked and what kind and extent of information should be required.

Undoubtedly, there are those who prefer that the feed regulatory official spell out exactly what is required - for example, a finite list of mycotoxins that must be tested for. But given today's knowledge, this approach does not appear to be feasible or reasonable.

As a matter of fact, there are various forms of evaluation guidelines published in the literature. It is up to the manufacturer to use these to develop his own testing protocol in order to arrive at scientific information that will satisfy himself, first and foremost, that his microbial biomass protein is safe and useful.

UNIFORMITY

An important characteristic that applies to both usefulness and safety is the uniformity or constancy of composition of the product. Uniformity must be such that a product can be relied upon to have the same properties from batch to batch. It is recognized that the information contained in a submission for pre-sale clearance/ registration relates to a product produced by pilot plants. However, it is incumbent upon the manufacturer to ensure that once the ingredient is produced on a commercial scale, the uniformity is maintained.

Once a new feed ingredient has been found acceptable under the Canada Feeds Act and Regulations and enters into use it is subject to periodic inspection, including sampling and analysis, by the Department of Agriculture. The objective is to determine whether or not the product continues to be true to the specifications on which its acceptance was based. Should the uniformity of the ingredient vary to the extent that its consistency is in question, regulatory action can be taken. Provisions are made in the Canada Feeds Regulations to revoke a registration if it is determined that, based on current information, the safety and efficacy of a feed for its intended purpose is no longer acceptable.

LABELLING

Some concerns were raised about the Regulations requiring complete disclosure of ingredients on a feed label. This could have an adverse effect on the acceptance of a microbial biomass protein if the purchasers react negatively to the inclusion of microorganisms and/or products thereof in feeds.

With Schedule IV being part of the current Canadian Feeds Regulations, there is no longer a requirement for commercial mixed feed manufacturers to disclose ingredient composition on feed labels (with the exception of a few types of specialty feeds). However, the manufacturer of a mixed feed is required, upon request of a feed purchaser, to provide a list of all the ingredients used in the manufacture of the feed.

GENERAL SUMMARY AND CONCLUSION

Four regulatory aspects of microbial biomass proteins as feed ingredients have been covered. The central theme is that the Canada Feeds Regulations exist to ensure the usefulness and the safety of livestock feeds and to prevent the entry of feed-borne residues into meat, milk and eggs. The feed regulatory official is charged with the responsibility of asking manufacturers of these novel protein ingredients questions on behalf of his clients, namely, the farmers and the mixed feed manufacturers.

Some may consider the regulatory body a hurdle to overcome. But for responsible and reputable feed manufacturers, addressing the issues raised is a necessary part of their product development process. The information they supply to the regulatory official is that which they must have in order to satisfy their customers and themselves.

The Feed and Fertilizer Division of the Canada Department of Agriculture is receptive to the potential introduction of microbial biomass proteins as feed ingredients in the Canadian marketplace. Hopefully, the regulatory approval process will enhance consumer acceptance of such novel products.

Appendix I

Safety Standards

19. (1) Subject to subsections (2) and (3), a feed shall not contain

.

.

.

(j) any material in quantities that could, when fed in proportions commonly used or as specified in the feeding directions, result in the production of an article of food that is prohibited from sale by virtue of section 4 of the *Food and Drugs Act;* or

(k) any material, other than those referred to in paragraphs (a) to (j), in quantities likely to be deleterious to livestock, when fed in proportions commonly used or as specified in the feeding directions.

Appendix II

Satisfactory Evidence

8. (1) In addition to the requirements set out in section 6, an applicant shall, if requested to do so by the Director, provide the Director with satisfactory evidence to permit an assessment or evaluation of the safety and efficacy of the feed in respect of livestock and its potential effect on humans and on the environment.

(2) The evidence referred to in subsection (1) shall, where appropriate, include the following information, descriptions and reports:

(a) the results of scientific investigations respecting
(i) the conditions and the prevalence of such conditions under which the feed would be efficacious for its intended purposes,
(ii) the safety of the feed in respect of the species of livestock for which it is intended and in respect of other species of livestock and humans who may be exposed to it,
(iii) suitable methodology for the detection of significant amounts of any ingredient, compound, substance or organism that is intentionally incorporated into the feed or that occurs as a contaminant of the feed,
(iv) harmful residues, if any,
(v) significant changes in the chemical or physical composition of livestock products produced when the feed is used, and
(vi) the stability of the feed under practical conditions of storage;

(b) a description of production methods including
(i) information with respect to actual formulas to be used in the manufacture of the feed,
(ii) information with respect to the type and capacity of the equipment to be used in the manufacture of the feed, and
(iii) information with respect to quality control procedures to assure uniformity of the mix and the lack of contamination of subsequent lots of feed manufactured in the same place; and

(c) reports of analysis for any specified nutrient or medicating ingredient that is required to be guaranteed in the feed conducted on at least three samples, each drawn from a different one-third of a single mix or batch of the feed.

(3) Where any investigation has been performed for the purpose of providing the evidence referred to in subsection (1), the applicant shall establish that

(a) the investigation was conducted or supervised by qualified research personnel;
(b) the investigation was designed to facilitate statistical analysis and the results of the investigation were analysed by appropriate statistical methods; and
(c) the investigation was conducted under conditions similar to those that may be expected to occur in Canada.

EVALUATING THE SAFETY AND NUTRITIONAL VALUE OF MICROBIAL BIOMASS

John N. Udall and Nevin S. Scrimshaw

Clinical Research Center
Massachusetts Institute of Technology
50 Ames Street
Cambridge, Massachusetts
U.S.A. 02114

INTRODUCTION

The search for an appropriate neutral name for microbial biomass produced on relatively pure carbohydrate or hydrocarbon substrates led to the introduction of the name Single Cell Protein as the title for a 1967 conference (1). The name, usually abbreviated to SCP, soon became the generic term throughout the world. When a conference was convened in Guatemala in 1978 on the bioconversion of organic residues (2), it was apparent that a name was needed to designate products based on microbial growth on vegetable and animal wastes in which the residues themselves are a significant part of the final product that has been enriched or transformed by the microbiological process applied. After considerable discussion, the name microbial biomass product (MBP) was proposed and we are pleased to see it used in the title of the meeting. It is equally applicable to wholly new applications of microbiology to the utilization of organic residues and to such traditional fermentation products as tempe, ontjom, and bongkrek (3).

Subsequent to the 1967 conference, the Protein Advisory Group of the UN system developed a series of guidelines, since revised and republished under United Nations University auspices, summarizing procedures for the preclinical (4) and clinical (5) evaluation of SCP and other new protein sources for human feeding and one for their use in animal feeding (6). The series is completed by one dealing specifically with the production of single cell protein for human consumption (7).

There need be no distinction between the application of the guidelines to SCP and to MBP for animal and human feeding. In both cases there must be assurance of the safety of the primary organism(s) and, where appropriate, absence of usual contaminating organisms. There must also be assurance that pathogenic species cannot be accidentally introduced without detection. As stressed in Guideline 6 (4), the substrate used for microbial conversion whether for SCP or MBP must be free of pathogens, and have acceptably low levels of heavy metals (Hg, Cu, Cd, Pb, Al, As), pesticide residues, drug residues, toxic metabolites and foreign bodies.

The assurances necessary for the application of microbiological techniques to organic residues are summarized in Table 1. Whether for animal or human use, they include the safety of the species of micro-organisms involved, the safety of the substrates, and satisfactory nutritional value. If for the feeding of the animals whose flesh or products are intended for food use, there must be the further assurance that the food products derived have not accumulated harmful substances or contain harmful metabolites.

TABLE 1: Assurances necessary for application of microgiological techniques to organic residues

For animal use:

(a) Safety of species
(b) Safety of substrates
(c) Safety of animal products
(d) Nutritional value

For human use:

(e) Lack of allergenicity
(f) Lack of mutagenicity/carcinogenicity
(g) Lack of teratogenicity
(h) Favorable organoleptic or functional characteristics
(i) Cultural acceptability

When direct human consumption is planned for the product, there must be the further assurance of low allergenicity; of the lack of mutagenicity, carcinogenicity or teratogenicity; of favorable organoleptic or functional characteristics; and of cultural acceptability.

GUIDELINES FOR CLINICAL TRIALS

While the revised PAG/UNU Guideline No. 6: **Preclinical Testing of Novel Sources of Food** applies to all novel protein sources, it has been most used as a standard for single-cell protein studies (4). It stresses that no new protein source should be submitted for clinical trials until it has been successfully fed to one rodent and one non-rodent species in short-term toxicological tests in which the experimental protein source is fed at the highest practical level. A full battery of biochemical tests are required, and when the animals die or are sacrificed, they should be examined for gross pathology, organ weight, and histopathological examination of the principal organs and tissues.

These short-term studies should be supplemented by examination of the new material using a battery of mutagenicity tests in both prokaryotic and eukaryotic systems. The short-term studies can then be extended to a multigeneration study that will assure satisfactory reproduction and lactation. The F_2 generation can be used for chronic toxicity and carcinogenicity studies and will provide information on possible teratological effects. All of the published clinical studies of new single-cell protein have been on materials that have passed through the preclinical testing requirements of Guideline 6.

There is an additional guideline, No. 12, specifically for the Production of Single-Cell Protein for Human Consumption. It lays down a series of additional requirements that are not all relevant to MBP. It states that the final product should contain no living cells derived from the fermentation process, a requirement that may or may not apply to MBP. Applicable to both SCP and MBP is the requirement that attention must be directed to the composition of the media to ensure that it does not contain chemical compounds that are regarded as health hazards. In order to ensure product quality and sufficient uniformity, it will be necessary in the production of MBP as in the case of SCP to pay attention to variables of the fermentation process as temperature, aeration and pH. Attention must also be paid to maintaining the strain or mix or organisms through appropriate microbiological examinations.

The original PAG guideline 7, Human Testing of Novel Foods, placed considerable stress on studies of nutritional value and described in detail both nitrogen balance in children and adults and growth studies in children for this purpose. In practice, the protein value of the various SCP's for human subjects can be predicted with sufficient accuracy by the use of amino acid scores from experimental animals. In fact, it can be assumed that any food that supplies sufficient protein with an acceptable amino acid pattern and a useful content of the essential nutrients will be of value to the consumer if it is well tolerated and acceptable. Accordingly, the revised guideline issued in 1983 places primary emphasis on tolerance studies in relatively large numbers of subjects to determine acceptability and the frequency of allergic and other undesirable reactions.

No amount of testing in experimental animals, including the testing of primates, can serve to predict the kinds of intolerance to SCP's that have been encountered in human subjects. In fact, no problems have been encountered with well processed SCP's prepared under carefully defined and maintained conditions. This is probably because the reactions observed appear to have been allergenic in nature and idiosyncratic in humans (8). The most common problems are gastrointestinal intolerance and skin rashes. Since these generally appear within two weeks, a thirty-day trial is sufficient to detect reactions that occur with enough frequency to be of concern.

Various allergic reactions may occur in some individuals after the consumption of almost any common food, especially if it is a protein source. It is the frequency of such reactions that must be evaluated rather than their expected absence. Moreover, when persons know they are, or believe they are, consuming a novel food, minor symptoms unrelated to the material ingested may be exaggerated or even imagined. For this reason, it is essential to feed a control group simultaneously.

Ideally, the trial is conducted so that individuals are randomly assigned to experimental and control groups, stratified by sex, in a double-blind cross-over design. Only after a low frequency of adverse symptoms is assured can additional trials be conducted without the necessity of a control group. It is inadvisable to test at a level markedly in excess of that at which the product is likely to be consumed. Even traditional foods tested at excessive levels might well give negative results.

In general, 25 and preferable 50 human subjects should be studied initially. Subjects should be chosen without reference to their personal and family history of allergies since it is essential to determine the prevalence of symptoms in the general population. The subjects should be in good health as determined by medical history and physical examination and routine blood and urine tests. The test material is to be consumed in addition to an *ad libitum* diet. Since this will be highly variable and may include meals or foods that in themselves cause symptoms, a control group of adequate size is essential. In addition, each subject should keep a diary of all significant departures from his/her usual diet, activities or the occurrence of symptoms. Even changes in mood, appetite, sleep patterns, libido, and other subjective reactions should be recorded. These diaries often prove of great value in the retrospective interpretation of the symptoms of a tolerance trial by making possible the detection of results that are unrelated to the test material or, conversely, identifying responses that were not obvious during the trial.

The guideline suggests that an appropriate study design is two four-week periods separated by a few days interval (5). In the first period, half of the subjects ingest daily a fixed amount of the experimental material and the remainder a control material that cannot be identified as different from that given the experimental group. If this is not possible, then groups should still not be allowed to know whether or not they are receiving the experimental material. After the initial four-week phase and an interval, the groups can be reversed if an undue number of reactions have not occurred. If there have been significant reactions, the trial must be terminated and the code broken. Not only must the trial be terminated at any time that an undue number of adverse reactions are recognized, but also for ethical reasons subjects must be allowed to leave the study at any time if they deisre. Administration of the material must be done for a minimum of five and preferably six or even seven days a week. Experience indicates that individuals not developing symptoms in the first 14 days are not likely to do so thereafter.

Any method of oral administration will suffice that facilitates a double-blind trial in which neither the subject nor the supervising physician knows which group is receiving the experimental material. Among the possibilities are concealing the novel food in cookies, cakes, and puddings. SCP's that are bland powders can often be tested by allowing the subjects to mix them into bouillon or any of a variety of fruit juices according to their individual preference. Subjects can be free to vary their choice of bouillon or fruit juice from day to day. For the study to be valid, there must be a similar control material for which tolerance is already well established. The level of feeding should be based on the intended level of use.

We have recently published applications of the PAG guidelines to the evaluation in human subjects of two filamentous microfungal products (9). One of these was *Fusarium graminearum* grown by continuous fermentation on a medium of commercial glucose syrup treated to reduce the nucleic acid content. The other was *Paecilomyces varioti*, produced on an energy substrate of the spent liquor of a sulfite wood pulping meal.

The tolerance of human subjects to the two microfungal food products was studied in separate double-blind cross-over studies. As an addition to the subject's usual diets, cookies with and without 20 g of a product from *F. graminearum* were fed to a group of 100 individuals daily. In a second study, cupcakes with and without 10 g of *P. varioti* were given daily to 50 individuals. Mild rashes possibly related to one of the microfungal food products occurred in two individuals fed *P. varioti*. Except for a decrease in serum cholesterol during the *F. graminearum* study, no significant changes were noted in 17 serum constituents. during nutritive value studies, digestibility, biological value, and net protein utilization were calculated for the two microfungal proteins and for milk. The values for milk were 95%, 85% and 80%, respectively. For *F. graminearum* were 78%, 84% and 65%, respectively. For *P. varioitii*. corresponding figures were 81%, 67% and 54%. On the basis of these results, both microfungal foods may be deemed safe for human consumption at the levels tested (9).

These successful studies contrast markedly with the problems we have had with some yeasts and bacteria (8). The principal problem encountered has been one of gastrointestinal or cutaneous allergies ranging from mild to severe and from rare to affecting almost all test subjects. In almost every case we have been successful in guiding the producer to processing changes that result in a satisfactory product, although sometimes at an unacceptable cost. There is no reason to doubt that as more and more MBP products are submitted to thorough testing, similar problems will be encountered, some due to the organism and others to the substrate. Since multiple organisms are usually involved in MBP's and the substrates are usually quite impure, the problems may well be greater than for SCP's. Nevertheless, the number of traditional MBP products used as food by various societies is highly encouraging in this regard.

REFERENCES

1) Scrimshaw NS. Introduction. In: Single-Cell Protein, RI Mateles and SR Tannenbaum (eds). Cambridge, MA: The MIT Press, 1968, pp. 3-7.

2) Bioconversion of Organic Residues for Rural Communities. Food and Nutrition Bulletin Supplement 2, November 1979. The United Nations University, Tokyo, Japan.

3) Indonesian tempe and related fermentations: Protein-rich vegetarian meat substitutes. In: Handbook of Indigenous Fermented Foods. KH Steinkraus (ed). New York: Marcel Dekker, Inc., 1983, pp. 1-94.

4) Protein-Calorie Advisory Group of the United Nations System (PAG)/United Nations University (UNU), Guideline No. 6, "Preclinical Testing of Novel Sources of Protein," United Nations, New York, printed in Food and Nutrition Bulletin, Vo. 5, No. 1, 1983.

5) PAG/UNU Guideline No. 7, "Human Testing of Supplementary Food Mixtures," United Nations, New York, printed in Food and Nutrition Bulletin, Vol 5, No. 1, 1983.

6) PAG/UNU Guideline No. 15, "Nutritional and Safety Aspects of Protein Sources for Animal Feedings," United Nations, New York, printed in Food and Nutrition Bulletin, Vol 5, No. 1, 1983.

7) PAG/UNU Guideline No. 12, "Production of Single-Cell Protein for Human Consumption," United Nations, New York, printed in Food and Nutrition Bulletin, Vol 5, No. 1, 1983.

8) Scrimshaw NS. Single-cell protein for human consumption - An overview. In: Single-Cell Protein II. ST Tannenbaum and DIC Wang (eds). Cambridge MA and London, England: MIT Press, 1975, p. 24.

9) Udall JN, Lo CN, Young VR, and Scrimshaw NS. The tolerance and nutritional value of two microfungal foods in human subjects. Am J Clin Nutr. 1984; 40:285-292.

GENETIC ASPECTS AND THE POSSIBLE HAZARD OF MYCOTOXINS IN FUNGAL PRODUCTS

J.D. Bu'Lock

Microbial Chemistry Laboratory
Victoria University of Manchester
Manchester M13 9PL
England

For the exploitation of the full versatility of fungi in biotechnology, the problem of possible product contamination with mycotoxins is an inconvenient but serious issue, particularly when fungi or fungal products are to be used as food ingredients or as agents in food processing. We must accept that it is the duty of regulatory bodies to ask difficult questions, and if the questions that emerge are ones we see as stupid, then it is up to us to develop the scientific grounds for re-phrasing those questions in more meaningful ways that are capable of being scientifically answered.

This involves inverting much of our approach to mycotoxins. Historically the study of mycotoxins has proceeded from the identification of a mycotoxicosis, through the species-labelling of an associated fungus, to the isolation and further study of a causative mycotoxin. When the biotechnological use of a specified fungus is proposed, the chain of reasoning is quite different. In terms of what has gone before, every link in the chain is newly questionable, and the range of consequent experiments and concepts will be correspondingly either different or inadequate.

It is one thing to identify by classical taxonomy the species designation of a fungus known to be producing mycotoxin. It is something quite different to proceed from the classic taxonomic designation of an isolate to deductions about the probability of its producing a mycotoxin and the circumstances under which this is most likely to occur. In attempting this we run into all sorts of fundamental uncertainties about what the taxonomic designation of a species really means, which are fundamentally questions about genetics and evolution.

Equally, it is an accepted thing to consider the effect of environmental parameters on mycotoxin production by a fungus known to be able to produce a mycotoxin; the problem has been widely studied and general principles are correspondingly accepted.

It is something different again to propose environmental conditions under which any possible ability to produce a mycotoxin will actually become manifest. Industrial methods of screening for new metabolites are desirable ways of reducing effort (1), and become essential investigative tools. Our existing knowledge of the phenotypic regulation of mycotoxin production becomes essential background information.

From such starting-points it is at least feasible to be able to develop answers to such questions as,

> "are there **any environmental conditions** under which the fungal isolate you are using produces a **significant** amount of any of the mycotoxins we would **expect** to **be formed** by an isolate thus **designate taxonomically**, and are any such conditions **guaranteed capable of avoidance** in the **whole series of circumstances** your **proposed process uses or necessarily implies may occur?**"

(the critical concepts in this question are in bold).

However, if the regulatory questions are taken further, we enter new areas of ignorance. Our ignorance of the taxonomic and phenotypic aspects of mycotoxin production is great, but our ignorance of the genetic aspects is almost total. Yet the regulatory questions are usually taken further:

> "assuming that your answer to the question as asked above is favourable; what answer can you give that also allows for genetic changes in the organism being used?"

In many cases, the underlying presumption is that since what we propose to use is a fungus (of some kind or other), there is necessarily a possibility that it is liable to produce a mycotoxin, and only some fortuitous circumstance, that a random genetic event might trigger, is preventing it. Even when the objection has such a trivial basis, we really need a much more certain knowledge than we currently have about fungal genetics, the genetics of secondary metabolism, and the relationship between secondary metabolism and generally-accepted taxonomic criteria, before we can give a satisfactory answer.

Today the beginnings of progress in these directions are becoming apparent. For example, the basis of the distinction between *Aspergillus flavus* and *A. oryzae* is now considerably clarified (2), but awaits interpretation in terms of modern genetic concepts. Here is at least one clear task. An issue that may be related is the question of how a mycotoxin-producing aspergillus fares in a competitive situation in mixed culture with a "domesticated" isolate; some answers to this question are also beginning to emerge (3).

As a result of a converging body of recent work, much of it unpublished, we are approaching a situation where the absence of mycotoxin production by an isolate that might "reasonably" be expected to produce a mycotoxin, and that has been tested for this possiblity under a fully-appropriate range of environmental circumstances, can be explained by any one of the following situations - for each of which appropriate tests can at least in principle be devised once they have been recognized:

(1) all or most of the genes for mycotoxin production (comprising both structural and regulatory genes) are absent (despite the species description). It should perhaps then be open to us to declare the isolate a separate species; such strains can only

acquire mycotoxin-producing ability through major genetic changes.

(2) some of the structural genes for mycotoxin production are absent or mutationally inactivated (note that such isolates should produce demonstrable levels of certain precursors under appropriate conditions). In simple genetic recombinants such as diploids or heterokaryons, or their further progeny, strains of this kind can complement other strains with different lesions in the overall pathway.

(3) a chromosomal regulatory gene for mycotoxin production has been mutated, transposed, or eliminated, leading to failure to activate the structural genes. Present experience is that such a change may be virtually irreversible or it may be rather labile, and only more detailed genetic analysis will differentiate between these situations, which for practical purposes is extremely different. However, even these inactive strains can probably acquire the activating element by recombination with other strains.

(4) an extra-chromosomal regulatory gene may be present that prevents activation of the chromosomal genes for mycotoxin production; in the examples known so far this can be a DNA element, in which case its relationship to the transposition-like events that seem to underly situations like (3) is in question, or it can be an RNA "virus". In either case its elimination will "unmask" mycotoxin production in a hitherto non-toxigenic strain; alternatively it might be used as an infectious detoxifier.

Based on our work on *Fusarium sp.* (4), on Bennett's work on *A. parasiticus* (5), and on some unpublished reports from other laboratories (6), examples of all four situations can be given. They must be interpreted as indicating directions and the beginnings of progress, rather than as establishing the final conclusions we must continue to work for.

ACKNOWLEDGEMENT

Our own work in the mycotoxin field was initiated with support from the Science & Engineering Research Council and R.H.M. Research Ltd., and is currently supported by DG.XII of the European Communities.

REFERENCES

1) R. Hutter, Design of culture media capable of provoking wide gene expression, in "Bioactive microbial metabolites I", eds. J.D. Bu'Lock, L.J. Nisbet and D.J. Winstanley, Academic Press, London, 1982, pp 51-70.

2) D.T. Wicklow, Adaptation in wild and domesticated yellow-green Aspergilli, in "Toxigenic fungi: their toxins and health hazards" eds. H. Kurata and Y. Ueno, Tokyo 1983, II-2, pp 78-86.

3) J.E. Smith, personal communication; *A. oryzae* markedly suppresses aflatoxin production in mixed culture with *A. flavus* under conditions that would normally favour aflatoxin production.

4) J.D. Bu'Lock, J.S. Duncan, J. Mooney and C. Wright, unpublished work; J.D. Bu'Lock and J.S. Duncan, Degeneration of zearalenone production in *Fusarium graminearum*, Experimental Mycology 9, in press (1985); J.D. Bu'Lock, Genetic aspects of mycotoxin formation, Hoechst Workshop Conference on Regulation of Secondary Metabolite Formation, 1985 (in press).

5) J. Bennett, personal communications; I am particularly grateful to her for continuing discussion of the problems I have raised here, and for her always constructive criticisms.

6) Thanks are particularly due to K. Esser for disclosing results on the "unmasking" of aflatoxin production in an *A. flavus* strain on elimination of a double-stranded RNA particle, and to M. Luckner for discussing the effects of extra-chromosomal DNA elements in controlling benzodiazepine formation in *Penicillium cyclopium*.

NUTRITIONAL EVALUATION OF INACTIVE DRIED FOOD YEAST PRODUCTS

G. Sarwar, R.W. Peace and H.G. Botting

Health and Welfare Canada, Bureau of Nutritional Sciences
Food Directorate, Health Protection Branch
Tunney's Pasture, Ottawa, Ontario, Canada K1A 0L2

ABSTRACT

Nucleic acid, dietary fiber, nutrient composition and protein quality of six commercial inactive dried food yeast products have been studied. The yeast products were produced from two organisms (*Candida utilis, Saccharomyces cerevisiae*) grown on different substrates (cane and/or beet molasses, calcium lignosulphate/wood sugars, sulphite waste liquor or ethyl alcohol). The yeast products contained: nucleic acids, 8.1-11.5%; true protein (amino acid N x 5.7), 32.6-43.6%; dietary fiber, 3.1-11.7%; total folacin, 22.4-75.2 μg/g; pantothenic acid, 106-291 μg/g; biotin, 0.85-1.85 μg/g; Ca, 0.2-4.6 mg/g; K, 19-33 mg/g; Fe, 10-120 μg/g; Zn, 70-100 μg/g; Mn, 0.2-153.0 μg/g. Relative net protein ratio (RNPR) values (casein + methionine = 100) of diets containing 8% protein from the yeast products were 40-69%. The yeast products were deficient in sulphur amino acids for rat growth. The corrected RNPR values (which included an adjustment for rat's higher requirement for methionine + cystine than that of human) for various yeast products were 60-100%. Species of yeasts, autolysis and/or growth on different substrates affected levels of nucleic acids, dietary fiber, most nutrients and protein quality.

The Health Protection Branch of the National Health and Welfare Department is responsible for evaluating safety and nutritional quality of new (yeast) products as food ingredients in Canada. Inactive dried food yeast products have been commercially promoted for various nutritional and functional applications (1-2). In order to facilitate evaluation of new inactive yeast products by the Health Protection Branch, chemical composition and nutritional quality of six commercial inactive dried food yeast products have been studied in the Nutrition Research Division (3-4). The results of these studies (3-4) and related observations form the basis of this paper.

DESCRIPTION OF YEAST PRODUCTS

Six commercial yeast products listed in Table 1 were studied. Two yeast products (A and B) were produced from *Saccharomyces cerevisiae* (grown on cane and/or beet molasses) while four products (C-F) were produced from *Candida utilis*. The *Candida utilis* products were grown on calcium lignosulphate/wood sugars (C), sulphite waste liquor (D) and ethyl alcohol (E and F). Products B (*Saccharomyces cerevisiae*) and F (*Candida utilis*) were autolysed extract and autolysate, respectively.

TABLE 1: Origin and substrate of inactive dried feed yeast products.[1]

Product	Organism	Substrate
A	*Saccharomyces cerevisiae*	Cane and/or beet molasses
B (Autolysed extract)	*Saccharomyces cerevisiae*	Cane and/or beet molasses
C	*Candida utilis*	Calcium lignosulphate/wood sugars
D	*Candida utilis*	Sulfite waste liquor
E	*Candida utilis*	Ethyl alcohol
F (Autolysed)	*Candida utilis*	Ethyl alcohol

[1] Sarwar et al. (3).

NUCLEIC ACIDS

A review of the literature revealed that suitable methods for the quantitative determination of purines and pyrimidines (or total nucleic acid nitrogen) in foods were scarce. Therefore, a simple and rapid HPLC method for the quantitative determination of purines and pyrimidines in yeasts and other food products was developed by Sarwar et al. (3). The hydrolysis of yeast nucleic acids into purines and pyrimidines was carried out with 11.6 N perchloric acid for 1 h at 100 °C, and 8 nucleobases were separated isocratically, using a Waters HPLC System, in about 12 minutes (Figure 1).

The six yeast products contained 8.1-11.5% total nucleic acids (Table 2). These differences could be due to differences in species of yeast and/or conditions of growth.

In 1975, the Protein Calorie Advisory Group recommended that the maximum safe limit of nucleic acids for adults from single cell proteins is 2 g/day and from all dietary sources is 4 g/day (5). The data presented in Table 2 suggest that intakes of less than 2 g of nucleic acids (or about 20 g of the yeast products) could be easily met by the nutritional and functional uses of the yeast products.

Increased intakes of nucleic acids should be avoided to prevent hyperuricemia and associated clinical conditions such as kidney stone formation or gout (5). Recently, it has been suggested that blood levels of uric acid may provide an assessment of risk of cardiovascular disease, independent of obesity (6).

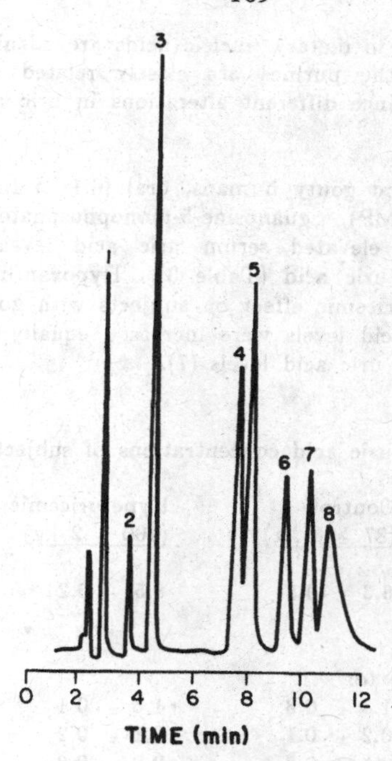

FIGURE 1: Chromatogram of standard nucleobases (absorbance at 254 nm, 0-0.2 AμFS). Peaks: 1 = cytosine, 2 = 5-methylcytosine, 3 = uracil, 4 = guanine, 5 = hypoxantine, 6 = Xanthine, 7 = thymine, 8 = adenine

TABLE 2: Purine, pyrimidine and nucleic acids data of inactive dried yeast products.[1]

Product	Adenine mg/100 g	Guanine mg/100 g	Cytosine mg/100 g	Uracil mg/100 g	Total nucleic acid N NAN, %	Nucleic acids NAN x 9, %
A	707[d]	624[f]	344[e]	457[e]	0.900[d]	8.10[d]
B	884[b]	925[b]	436[c]	631[c]	1.209[b]	10.87[b]
C	908[b]	943[a]	524[a]	680[b]	1.276[a]	11.47[a]
D	944[a]	912[c]	499[a]	711[a]	1.279[a]	11.50[a]
E	947[a]	801[d]	469[b]	670[b]	1.207[b]	10.86[b]
F	781[c]	656[e]	399[d]	554[d]	1.000[c]	9.00[c]

[1] Sarwar et al. (3).
[a-f] Means within the same column bearing different letters are significantly ($P < 0.05$) different.

The purines that contribute to dietary nucleic acids are adenine, guanine, xanthine and hypoxanthine. Although the purines are closely related compounds, they are metabolised differently and produce different alterations in uric acid metabolism when fed to humans or animals (7-8).

In normal, hyperuricemic and gouty humans, oral (0.1 m mole/kg) hypoxanthine, adenosine-5-monophosphate (AMP), guanosine-5-monophosphate (GMP), inosine-5-phosphate (IMP) and adenine elevated serum uric acid levels while guanine and xanthine did not affect serum uric acid (Table 3). Hypoxanthine, AMP, GMP and IMP produced a greater hyperuricemic effect on subjects with gout compared to other subject groups. Urinary uric acid levels were increased equally by all purines except for guanine which did not alter uric acid levels (7).

TABLE 3: Changes in serum uric acid concentrations of subjects fed oral purines.[1]

Subject Group	Control (87 ± 6 kg)	Hyperuricemic (100 ± 2 kg)	Gouty (87 ± 3 kg)
Fasting serum uric acid (mg/100 ml)	6.3 ± 0.5	8.5 ± 0.2	8.3 ± 0.7
Changes in serum uric acid due to:			
Adenine	+1.8 ± 0.3	+1.6 ± 0.1	+2.0 ± 0.2
Guanine	-0.2 ± 0.1	+0.1 ± 0.2	+0.2 ± 0.3
Hypoxanthine	+2.4 ± 0.2	+2.8 ± 0.2	+4.1 ± 0.4
Xanthine	+0.7 ± 0.1	-	+0.7 ± 0.1
AMP	+2.2 ± 0.3	+1.9 ± 0.1	+4.2 ± 0.4
GMP	+2.3 ± 0.6	+2.2 ± 0.2	+3.3 ± 0.2
IMP	+1.5 ± 0.1	+1.7 ± 0.2	+3.2 ± 0.5

[1] Abstracted from Clifford et al. (7).

Addition of 0.75% of adenine to rat diets has been shown to result in reduced growth, altered hepatic enzyme activities and changes in purine excretion patterns in urine (8). However, guanine, xanthine and hypoxanthine did not affect rat growth or patterns of purines excreted in the urine.

The differences in metabolic effects of individual purines would suggest that the Protein Advisory Group's recommendation of maximum safe limit of nucleic acids may have to be revised to include safe limits of certain individual purines. Foods high in total nucleic acids may be low in uricogenic purines (such as hypoxanthine and adenine) and vice versa (8). Information on the purine contents of foods and additional studies on digestion, adsorption and metabolic effects of purines are needed for possible revision of the PAG Guideline.

NITROGEN AND PROTEIN

The yeast products contained 7.44-9.32% total nitrogen (Table 4). Nucleic acid and amino acid nitrogen constituted 12-16 and 78-84% of the total nitrogen in the yeast products, respectively, resulting in total nitrogen recovery of 94-98% (Tables 2 and 4). The six yeast products contained 32.57-43.63% true protein (amino acid residue), calculated as the sum of amino acids minus the elements of water (Table 4). The *C. utilis* products (C-F) contained more protein than the *S. cerevisiae* products (A-B). Among the *C. utilis* products (C-F), differences in protein content could be due to growth on different substrates. The lower levels of protein in product B (*S. cerevisiae*, autolysed extract) compared to A (*S. cerevisiae*), and in F (*C. utilis*, autolysed) compared to E (*C. utilis*) suggested that autolysis had decreased the contents of protein in both the species of yeast.

TABLE 4: Nitrogen and protein contents and nitrogen-to-protein conversion factors.[1]

Product	Total nitrogen %	Amino acid nitrogen %	Amino acid residue (true protein, %)	Nitrogen-to protein conversion factor
A	7.44	6.26	36.36	5.81
B	7.42	5.79	32.57	5.62
C	8.72	7.11	40.63	5.71
D	9.32	7.65	43.63	5.70
E	8.43	7.07	40.55	5.73
F	8.07	6.57	38.08	5.80

[1] Sarwar et al. (3).

True nitrogen-to-protein conversion factors were calculated by dividing amino acid residue by amino acid nitrogen (Table 4). The factors for various yeast products ranged from 5.62-5.81 (with an average of 5.73). These factors are markedly lower than the commonly used nitrogen-to-protein conversion factor of 6.25.

DIETARY FIBER

The dietary fiber contents of all the yeast products except product B were similar (15.42-17.79%) (Table 5). The comparison of product E (*C. utilis*) and product F (*C. utilis*, autolysed) revealed that autolysis reduced the level of cellulose by about 50% but had little effect on the level of noncellulosic polysaccharides. Product B (*S. cerevisiae*, autolysed extract) contained the lowest level of dietary fibre (Table 5). The preparation of product B which involved autolysis followed by centrifugation, may be responsible for eliminating most of the fiber polysaccharides.

TABLE 5: Data on dietary fiber contents of inactive dried yeast products.[1]

Product	Cellulose	Non-cellulosic polysaccharides	Uronic acid	Total fiber
A	4.84	11.45	tr	16.29
B	0.07	2.99	tr	3.06
C	4.82	12.97	tr	17.79
D	4.23	11.19	tr	15.42
E	5.49	12.24	tr	17.73
F	2.47	12.95	0.03	15.45

[1] Sarwar et al. (3).

ASH AND MINERAL NUTRIENTS

Ash contents of all the yeast products except product B ranged from 5.5-7.7% (Table 6). Due to the addition of 37% salt by the manufacturer, ash content of product B was very high. Wide variations in contents of Ca and Mn of various yeast products were noted (Table 6). Product C (*C. utilis*, grown on calcium lignosulphate/wood sugars) and product D (*C. utilis*, grown on sulphite waste liquor) contained the highest levels of Ca and Mn, respectively. These two minerals could have been picked up from the substrates used which may contain high levels of these minerals.

TABLE 6: Data on ash and some mineral nutrients in inactive dried yeast products.[1]

Product	Ash	Ca	K	Fe	Zn	Mn
	%	mg/g		μg/g		
A	5.8	0.60	28	70	70	4
B	42.8	0.18	18	10	80	0.2
C	5.9	4.64	22	120	100	27
D	5.0	0.89	19	100	80	153
E	7.5	0.05	33	110	70	5
F	7.7	0.09	33	50	70	5

[1] Sarwar et al. (3).

VITAMINS

The *S. cerevisiae* products (A and B) were lower in total folacin and pantothenic acid than the *C. utilis* products (C-F) but product A contained more biotin than other products (Table 7). Comparisons of product A with B and of E with F suggested that autolysis reduced initial amounts of the three vitamins. Among the *C.*

utilis products, substrates influenced levels of total folacin (52.8-75.2 μg/g) and pantothenic acid (151-291 μg/g) (Table 7). The biotin contents of the various products ranged from 0.84-1.85 μg/g.

TABLE 7: Data on folacin, pantothenic acid and biotin contents of inactive dried yeast products[1]

Product	Vitamin content, μg/g of product		
	Total folacin	Pantothenic acid	Biotin
A	36.2[c]	127[cd]	1.85[a]
B	22.4[d]	106[d]	1.34[b]
C	75.2[a]	151[bc]	1.30[b]
D	53.4[b]	126[cd]	0.84[c]
E	52.8[b]	291[a]	0.85[c]
F	26.5[d]	173[b]	1.18[b]

[1] Sarwar et al. (3).

[a-d] Means within the same column bearing different letters are significantly ($P < 0.05$) different.

PROTEIN QUALITY

Protein efficiency ratio (PER) and net protein ratio (NPR) values of diets containing 8% protein from casein + 0.2% L-methionine (control) or different yeast products are shown in Table 8. Product A (*S. cerevisiae*) had higher PER and NPR values while products F and E (*C. utilis*, grown on ethyl alcohol) had lower PER and NPR values than other yeast products.

TABLE 8: Data on protein efficiency ratio (PER), net protein ratio (NPR) and relative NPR (RNPR) values and sulphur amino acids.[1]

Diets	PER	NPR	RNPR	Met + Cys (g/100 g true protein)
Casein + Met control	4.43[a]	5.64[a]	100[a]	5.99
Yeast A	2.63[b]	3.91[b]	69[b]	3.50
Yeast B	1.97[cd]	3.24[c]	57[c]	2.96
Yeast C	2.09[c]	3.33[c]	59[c]	3.18
Yeast D	2.10[c]	3.25[c]	58[c]	3.36
Yeast E	1.83[d]	2.73[d]	48[d]	2.65
Yeast F	1.01[e]	2.24[e]	40[e]	2.56

[1] Sarwar et al. (4).

[a-e] Means within the same column bearing different letters are significantly ($P < 0.05$) different.

Recent collaborative projects (9-11) have demonstrated that relative NPR (RNPR) is the most appropriate rat growth assay for routine use in assessment of protein quality of foods. Therefore, the RNPR values of the yeast products were also calculated (Table 8). The values varied from 40-69%. As noted in the case of PER and NPR, yeast A had higher RNPR value while yeasts F and E had lower RNPR values than other yeasts (Table 7). These differences in protein quality of yeast products could be due to differences in their contents of methionine + cystine (Table 8), the first limiting amino acids (4).

There is general agreement in the scientific community that the requirements of rats for methionine + cystine are much higher than those of humans (12). Therefore, any rat growth assay including RNPR will underestimate the protein quality for humans of any protein product deficient in sulphur amino acids.

Essential amino acid requirements of human and rat were recently compared by Sarwar et al. (Table 9). The differences between the requirements of all essential amino acids except methionine + cystine were small. The rat's requirement for methionine + cystine was about 50% higher than that of human (Table 9). Based on this difference in methionine + cystine requirements, a correction factor of 1.5 was used to correct RNPR values of those protein products which were deficient in sulphur amino acids for rat growth.

TABLE 9: Comparison of human and rat requirements for essential amino acids.[1]

Amino Acid (% of protein)	Rat	Human	Rat/Human ratio
Isoleucine	4.17	3.65	1.14
Leucine	6.25	7.15	0.87
Lysine	5.84	5.75	1.01
Methionine + cystine	4.00	2.65	1.51
Phenylalanine + tyrosine	6.67	7.10	0.94
Threonine	4.17	3.60	1.16
Tryptophan	1.25	1.17	1.07
Valine	5.00	4.30	1.16

[1] Sarwar et al. (12).

Since the yeast products tested were first limiting in sulphur amino acids for rat growth (4), corrected RNPR (CRNPR) values were calculated for these products. The CRNPR values for yeast A, B, C, D, E and F were 100, 85, 88, 87, 72 and 60%, respectively.

REFERENCES

1) Protein Update 78. 1978. Part II. Guide to protein products and companies. Food Processing, July, 44.

2) Litchfield, J.H. 1983. Single-cell proteins. Science 219:740.

3) Sarwar, G., Shah, B.G. Mongeau, R. and Hoppner, K. 1985. Nucleic acid, fibre and nutrient composition of inactive dried food yeast products. J. Food. Sci. 50:353.

4) Sarwar, G., Peace, R.W. and Botting, H.G. 1986. Metabolic effects of feeding inactive dried food yeast products to rats. J. Nutr. (Submitted for publication).

5) PAG. 1975. PAG and hoc working group on clinical evaluation and acceptable nucleic acid levels of SCP for human consumption. Protein Calorie Advisory Group of the United Nations System (PAG) Bulletin Volume V (3):17.

6) Fessel, W.J. 1980. High uric acid as an indicator of cardiovascular disease. Am. J. Med. 68:401.

7) Clifford, A.J., Riumallo, J.A., Young, V.R. and Scrimshaw, N.S. 1976. Effect of oral purines on serum and urinary uric acid of normal, hyperuricemic and gouty humans. J. Nutr. 106:428.

8) Clifford, A.J. and Story, D.L. 1976. Level of purine in food and their metabolic effects in rats. J. Nutr. 106:435.

9) Sarwar, G., Blair, R., Friedman, M., Gumbmann, M.R., Hackler, L.R., Pellett, P.L. and Smith, T.K. 1984. Inter- and intra-laboratory variability in rat growth assays for estimating protein quality of foods. J. Assoc. Off. Anal. Chem. 67:976.

10) McLaughlan, J.M., Anderson, G.H., Hackler, L.R., Hill, D.C., Jansen, G.R., Keith, M.O., Sarwar, G. and Sosulski, F.W. 1980. Assessment of rat growth methods for estimating protein quality: Interlaboratory Study. J. Assoc. Off. Anal. Chem. 63:462.

11) Sarwar, G., Blair, R., Friedman, M., Gumbmann, M.R., Hackler, L.R., Pellett, P.L. and Smith, T.K. 1985. Comparison of interlaboratory variation in amino acid analysis and rat growth assays for evaluating protein quality. J. Assoc. Off. Anal. Chem. 68:52.

12) Sarwar, G., Peace, R.W. and Botting, H.G. 1985. Corrected relative net protein ratio (CRNPR) method based on differences in rat and human requirements for sulphur amino acids. J. Assoc. Off. Anal. Chem. 68:689.

THERMOPHILIC AND THERMOTOLERANT FUNGI FOR INDUSTRIAL APPLICATIONS: MYCOLOGY ASPECTS

S. Udagawa*, S. Sekita*, and S. Natori**

* National Institute of Hygienic Sciences, Tokyo, Japan

** Meiji College of Pharmacy, Tokyo, Japan

Thermophilic and thermotolerant fungi are wide-spread and are potentially useful in a variety of industrial applications, including the rapid degradation and recycling of organic wastes, because of their luxuriant growth and cellulolytic activity. Prior to the developmental research, however, attention needs to be given to their detrimental activities. Potential health hazards of the fungi are divided into two parts: first, human and animal mycoses, and second, mycotoxicoses. Attention is focused on four species in this presentation.

Thermophilic fungi are those with a maximum temperature for growth at or above 50°C and a minimum temperature for growth at or above 20°C. Early work on the taxonomy of these fungi has been well documented by Cooney and Emerson (1964). In general, thermotolerant fungi have maxima near 50°C but minima well below 20°C. Here I should like to assume a broader concept for thermotolerant such as a maximum for growth at 40 to 45°C.

Thermophilic and thermotolerant fungi grow well at the temperature of the human and warm-blooded animal body, and some species are the causal agents of disease in humans and domesticated animals (Table 1). All of them are opportunistic and the immune response in a normal human usually prevents serious infection. Aspergillosis due to *A. fumigatus* has a world-wide distribution, and mucormycosis due to Mucorales is also common in Japan. But infections by the other fungi listed have rarely been reported in Japan or elsewhere.

Thermoascus aegyptiacus is possibly an intermediate form between *T. thermophilus* and *T. crustaceus*. It was isolated from marine sludge from the Suez Canal. Table 2 shows that ascospores of *T. aegyptiacus* are quite similar to those of *T. crustaceus*, both having fine ornamentation. These are almost smooth under a light microscope. On the other hand, ascospores of *T. crustaceus* are spinulose, even seen by a light microscope. An anamorph of this fungus is Paecilomyces which is very similar to *T. crustaceus.*

Sell et al. (1983) reported three unusual isolates of *Thermoascus crustaceus* which were obtained from three AIDS monocyte cultures. Their preliminary data showed that the mycelial extracts from all three isolates contained a cyclosporin-like compound. Furthermore, cell-free culture filtrates of the fungus were found to contain nontoxic immunosuppressive material as determined by lymphocyte bioassays. Kwon-Chung et al. (1984) described these unusual isolates of *T. crustaceus* in more detail. They pointed out that the isolates are almost idential to *T. aegyptiacus*

The second thermophilic fungus in this contribution is *Dactylaria gallopava.* Next to the fungi causing aspergillosis and mucormycosis, this fungus is of greatest interest due to its relative significance as a causal agent of human and animal diseases. In 1983, Prof. Fukushiro, Department of Dermatology, Kanazawa Medical University, found a hyphomycete as a case of opportunistic subcutaneous infection in a female patient suffering from acute leukemia. She died later because of the leukemia. One of us (Udagawa, 1980) identified this fungus as *D. gallopava;* the first such isolate in Japan. It grew very slowly at 51 °C and 18 °C. It grew most rapidly at temperatures between 35 °C and 40 °C with abundant sporulation.

This fungus was first reported as the causal agent of encephalitis in turkeys in South Carolina. An outbreak in 1962 was reported in a flock of 4,000 birds, of which 600 developed symptoms of the disease and 400 died. The source of the fungus in this outbreak was not known, but old sawdust litter used for bedding in the turkey houses may have been heavily contaminated with this fungus. A second report was made in 1967 in Australian chickens. After 10 years' absence, a second case in the USA was reported from South Carolina. In this outbreak, flock mortality eventually reached about 20%. Experimental infection was established by artificial inoculation of a spore suspension of *D. gallopava,* and revealed similar symptoms. This disease is widely spread over Georgia, Maryland, Indiana, and possibly in Scotland. Epidemiological interest in this disease resulted in the discovery by Tansey and Brock (1973) of its saprophytic phase in natural thermal and man-made heated habitats. They isolated it from effluents of acid hot springs and acid thermal soils in Yellowstone National Park and from self-heated coal waste piles in Pennsylvania. The occurrence of *D. gallopava* in coal waste piles has been reported also in England. Later Tansey and Fliermans (1979) described the growth, occurrence, and distribution of *D. gallopava* in thermal effluents from nuclear production reactors at the Savannah River Plant in South Carolina. Large populations of the fungus occurred in microbial mats, foam, and soil at the edges of the reactor cooling-water effluents.

The patient reported from Japan had no known exposure to a geothermal environment. There are no reports of infected chicks and turkeys in Japan but it is potentially very dangerous for poultry farms.

Susceptibility tests of *D. gallopava* were performed in liquid and solid media at 37 °C. Table 3 summarizes results on the sensitivity of three strains of *D. gallopava* to antifungal drugs *in vitro.* Both pathogenic strains were relatively more resistant than the non-pathogenic one, to all antifungal drugs. *Dactylaria gallopava* was transferred to the genus Ochroconis by de Hoog (1983) but is still commonly known by the old name Dactylaria.

The third fungus is a new species of Scytalidium, designated *S. japonicum*. Last year, Dr. Sato and his co-workers in the National Institute of Animal and Health, Japan, found two cases of bronchomycosis in cattle that involved pulmonary emphysema. An unusual strain of Scytalidium was isolated from the cattle-lung lesion and identified as a new species of the genus.

According to Sigler and Carmichaels (1976), there are seven known species of Scytalidium. Six additional species have been reported up to the present time, and three of them are characterized by their thermophilous nature. This new species differs from all other known species by its very large hyaline arthroconidia, the regular presence of dematiaceous arthroconidia, and the rather thermotolerant nature. Table 4, shows the morphological differences between *S. japonicum* and the three known species described as pathogenic. Scant attention has been paid to Scytalidium infections in medical and veterinary mycological fields, but according to present knowledge, there are three species that have been islated from clinical specimens and that are apparently causal in infections.

The Scytalidium anamorph of *Hendersonula toruloidea* is the most common species represented, recovered from infections of the skin and nails in humans. At first, most cases were reported by Gentles and Evans (1970), and Campbell et al. (1973) in former natives of tropical countries now residing in the United Kingdom. Later, Mariat et al. (1978) described a verrucose dermatitis of the human face, accompanied by onychomycosis, from Algeria. The fungus was isolated from facial lesions and affected nails. *Hendersonula toruloidea* (pycnidial form) is a widespread wound parasite of fruit trees of citrus and stone fruits and walnuts in tropical areas. The second species of pathogenic Scytalidium is *S. hyalinum*. The first reported case was in patients of Jamaican and West African origin with skin and nail infections. Campbell and Mulder (1977) reported that this fungus was actively invading and growing in diseased tissues. Recently Moore et al. (1984) demonstrated an *S. jyalinum* infection of human skin from Spain. It is surprising that all previous strains of this fungus were of clinical origin.

In 1983, a case of subcutaneous mycosis caused by *S. lignicola* was reported by Dickinson et al. from Florida. Previously, Dixon et al. (1980) had described the isolation of *S. lignicola* from sawdust samples collected in Virginia. Using hamster inoculation techniques, they confirmed recovery of the isolate of *S. lignicola* from homogenized tissue of hamster spleen 6-7 weeks post inoculation and concluded that the persistence of the fungus in an animal for that length of time could be considered to indicate potential pathogenicity.

Although not reported from a human infection, this was the first case of Scytalidium-mycosis characterized by systemic infection. The species involved in opportunistic infection grow well or preferentially at 37 ° C. Perhaps this agent has a greater capacity to adapt to a tissue environment than related species, and a greater potential virulence.

Despite the importance of thermophilic and thermotolerant fungi as agents of spoilage in agricultural commodities, little attention has been paid to their mycotoxigenic activity. According to the results of our mycotoxin screening survey of Chaetomium species, two thermotolerant species, *C. virescens* and *C. gracile* produce known or unknown toxic metabolites. The former species is specially noteworthy because of its production of sterigmatocystin. In our opinion (Sekita et al, 1980), *Chaetomium thielavioideum* and *C. cellulolyticum* are essentially identical with *Achaetomiella virescens.*

In the fields of medical mycology and mycotoxicology, the role of thermophilic fungi is often under-estimated. The above findings have added much to our knowledge in these fields. The possibility that thermophilic fungi are pathogenic or mycotoxin producers is significant to human and animal health. They are clearly not major pathogens but further studies are now necessary to determine the role these fungi play with respect to human and animal disease. Finally, we can point out that the utilization of thermophilic and thermotolerant fungi in the microbial industry requires careful testing for their detrimental activities.

REFERENCES

1) Campbell, C. K (1971). Studies on *Hendersonula toruloidea* isolated from human skin and nail. Sabouraudia 12: 150 - 156.

2) Campbell, C. K., A. Kurwa, A-H. M. Abdel-Aziz, and C. Hodgson (1973). Fungal infection of skin and nails by *Hendersonula toruloidea* Br. J. Dermatol. 89: 45 - 52.

3) Campbell, C. K., and J. L. Mulder (1977). Skin and nail infection by *Scytalidium hyalinum* sp. nov. Sabouraudia 15: 161 - 166.

4) Cooney, D. G., and R. Emerson (1964). Thermophilic Fungi. An account of their biology, activities, and classification. W. H. Freeman, San Francisco.

5) Dickinson, G. M., T. J. Cleary, T. Sanderson, and M. R. McGinnis (1983). First case of subcutaneous phaeophyphomycosis caused by *Scytalidium lignicola* in a human. J. Clinical Microbiol. 17: 155 - 158.

6) Dixon, D. M., H. J. Shadomy, and S. Shadomy (1980). Dematiaceous fungal pathogens isolated from nature. Mycopathologia 70: 153 - 161.

7) Gentles, J. C., and E. G. V. Evans (1970). Infection of the feet and nails with *Hendersonula toruloidea.* Sabouraudia 8: 72 - 75.

8) de Hoog, G. S (1983). On the potentially pathogenic dematiaceous Hyphomycetes, p. 149 - 216. In D. H. Howard (ed.), Fungi Pathogenic for Humans and Animals, Part A. Biology. Marcel Dekker, New York.

9) Kwon-Chung, K. J., T. Folks, and K. W. Sell (1984). Unusual isolates of *Thermoascus crustaceus* from three monocyte cultures of AIDS patients. Mycologia 76: 375 - 379.

10) Mariat, F., B. Liautaud, Marlene Liautaud, and F-G Marill (1978). *Hendersonula toruloidea,* agent d'une dermatite verruqueuse mycosique observée en Algerie. Sabouraudia 16: 133-140.

11) Moore, M.K., A. Del Palacio-Hernanz, and S. Lopez-Gomez (1984). *Scytalidium hyalinum* infection diagnosed in Spain. Sabouraudia 22: 243-245.

12) Sell, K.W., T. Folks, K.J. Kwon-Chung, J. Coligan, and W.L. Maloy (1983). Cyclosporin immunosuppression as the possible etiology of AIDS. New England J. Med. 309: 1065.

13) Sigler, L., and J.W. Carmichael (1976). Taxonomy of Malbranchea and some other hyphomycetes with arthroconidia. Mycotaxon 4: 349-488.

14) Tansey, M.R., and T.D. Brock (1973). *Dactylaria gallopava,* a cause of avian encephalitis, in hot spring effluents, thermal soils and self-heated coal waste piles. Nature, Lond. 242: 202-203.

15) Tansey, M.R., and C.B. Fliermans (1978). Pathogenic species of thermophilic and thermotolerant fungi in reactor effluents of the Savannah River Plant, p. 663-690. In J.H. Thorp and J.W. Gibbons (eds.), Energy and Environmental Stress in Aquatic Systems. Nat. Tech. Info. Serv. (USA), CONF. 77114.

16) Udagawa, S (1980). New or noteworthy Ascomycetes from Southeast Asian soil I. Trans. mycol. Soc. Japan 21: 17-34.

TABLE 1: Pathogenic species of thermophilic and thermotolerant fungi

Fungus	Pathogenicity	Occurrence in Japan
Absidia corymbifera	Mucormycosis	+
Rhizomucor miehei	Mucormycosis	+ (saprophytic)
Rhizomucor pusillus	Mucormycosis	+
Rhizopus microsporus	Mucormycosis	+ (animal)
Rhizopus rhizopodiformis	Mucormycosis	+ (saprophytic)
Emericella nidulans*	Aspergillosis	+
Aspergillus fumigatus	Aspergillosis	+ (common)
Aspergillus terreus	Aspergillosis	+
Thermoascus crustaceus	Rare	+ (saprophytic)
Coprinus cinereus	Rare	+ (saprophytic)
Phanerochaete chrysosporium**	Single report	
Ochroconis (Dactylaria) gallopava	Dactylariosis	+ (first case)
Scytalidium spp.	Rare	+ (animal)
Paecilomyces variotii	Rare	+
Thermomyces lanuginosus	Single report	+

*Anamorph: Aspergillus; ** Anamorph: Sporotrichum pruinosum.

TABLE 2: Comparative morphology of *Thermoascus* spp.

Character	T. aegyptiacus (1983)	Atypical T. crustaceus (1984)*
Ascomata	Orange brown, subglobose to irregular-shaped, 250-550 μm diam.	Brick to cinnamon, globose to oval, 150-300 μm diam.
Ascospores	Nearly smooth, ellipsoid, 7-8 x 4-5.5 μm	Nearly smooth, ellipsoid, 7 x 5 μm (averaging)
Anamorph	Paecilomyces	Paecilomyces
Phialides	Simple at 37°C, highly branched at 43°C	Simple at 25°C, highly branched at 37-45°C
Growth temperature	25-55°C	25-55°C
Opt. temp. for ascomata	30-40°C	37-40°C
Opt. temp. for conidia	45°C	40-45°C

* Sell et al. (16) and Kwon-Chung et al. (11).

TABLE 3: Sensitivity of *Ochroconis (Dactylaria) gallopava* to antifungal drugs

Strain*	MIC-values <u>in vitro</u> (µg/ml)**				
	AMPH	MCZ	KCZ	5-FC	GRF
Pathogenic					
NHL 2916	3.125	6.25	12.5	>100	>100
ATCC 26822	3.125	6.25	12.5	>100	>100
Non-pathogenic					
ATCC 26841	0.4	3.125	6.25	12.5	>100

* NHL 2916: from subcutaneous abscess tissue in human, Japan.
ATCC 26822: from brain abscess in turkey poult, USA.
ATCC 26841: from self-heated coal mine waste, USA.
Starting inoculum of ca 10^6 conidia/ml, respectively

** MIC = Minimum inhibitory concentration.
Antifungal drugs: Amphotericin B (AMPH), Miconazole (MCZ), Ketoconazole (KCZ), 5-Fluorocytosine (5-FC); Griseofulvin (GRF).

TABLE 4: Comparison of cultural and morphological characters between *Scytalidium japonicum* and other pathogenic Scytalidium

Character	S. japonicum NHL 2954	S. lignicola	S. hyalinum	Hendersonula toruloidea
Colony	olive gray	tan, gray or black	white	dark gray-black
Pycnidia	not produced	not produced	not produced	produced
Hyaline conidia (arthrospores)	9-22(-26) x 4.5-6.5 µm, 18-35 µm long (1-septate)	4.5-8 x 2 µm	5-12 x 2.5-3.5 µm or 4-6 µm diam	not produced
Dematiaceous arthroconidia (chlamydospores)	5.5-18 x 4-6 µm or 7-10 µm diam	7.5-12 x 4-7 µm, 12-17 µm long (1-septate)	not produced	6.5-15 x 3.5-5 µm
Growth at 37°C	+	−	+	+
Infected region	lung	skin	skin and nails	skin and nails
Reference	Authors	Dickinson et al. (1983)	Campbell and Mulder (1977)	Campbell (1974)

THERMOPHILIC AND THERMOTOLERANT FUNGI FOR INDUSTRIAL APPLICATIONS: CHEMICAL ASPECTS OF MYCOTOXINS

S. Sekita,* S. Natori**, and S. Udagawa*

*National Institute of Hygienic Sciences, Tokyo, Japan

**Meiji College of Pharmacy, Tokyo, Japan

The mycological aspects of thermophilic and thermotolerant fungi were described by Udagawa et al. elsewhere. Among these fungi, some Chaetomium species are being considered for the production of microbial proteins. In this paper, our recent work on the metabolites of *Chaetomium spp.* will be presented.

Chaetomium is one of the extremely familiar genera of Ascomycetes encountered on various agricultural commodities. The genus contains 200 or more species. Many of the species are of worldwide distribution and are common in soil containing plant debris. Because of their strong cellulolytic activity, several members of the genus are found as contaminants of straw, roots, tubers, seeds, grains, and other plant material. They may cause a biodeterioration of foods, feeds, and raw materials of pharmaceutical preparations.

Production by the genus of some secondary metabolites such as chaetomin (23) showing antibiotic properties had been known for a long time, but systematic surveys on mycotoxin production by Chaetomium had not been conducted until recently. Stimulated by the discovery of chaetoglobosins (11), a novel type of cytochalasans, we commenced a project to screen the metabolites of the genus in 1977. We examined about 120 strains belonging to about 70 species of Chaetomium using thin-layer chromatography and cytotoxicity testing on HeLa cells (12, 20). At nearly the same time, Canadian workers conducted a survey on 100 isolates of Chaetomium (1).

CHEMISTRY AND TOXICOLOGY OF CHAETOMIUM MYCOTOXINS

The mycotoxins produced by Chaetomium may be classified into six groups by their chemical structures as shown in Table 1. From a biogenesis point of view, the first three will be considered to be polyketides (acetogenins), formed from acetate-malonates. The other three groups are different each other, but all contain indole units derived from tryptophan.

TABLE 1: Chaetomium Mycotoxins

A) Polyketides (from Acetate-Malonate)
   1) Chaetochromin
   2) Sterigmatocystin, O-Methylsterigmatocystin
   3) Mollicellins

B) Indole Derivatives (from Tryptophan)
   1) Biindolylbenzoquinones
   2) Chaetoglobosins (10-indol-3-yl-<13>cytochalasans)
   3) Epipolythiodioxopiperazines

Chaetochromin (Figure 1) was originally isolated from *Chaetomium virescens* in our laboratory (13). Later the same compound was also isolated from *Chaetomium caprinum, C. gracile* and *C. tetrasporum* (12). The chemical structure is a dimer of naphthodihydropyrone (benzochromane) corresponding to a dimethyl derivative of cephalochromin (19), first isolated from a *Cephalosporium* sp.. Chaetochromin induces delayed liver injuries, atrophy of lymphatic tissues, and bone marrow aplasia, in mice (9). The compound also exhibits teratogenicity such as exencephaly (5).

The second group of compounds consists of sterigmatocystin and *o*-methylsterigmatocystin (Figure 1). They are produced by *C. virescens* and *C. udagawae* (12, 20). Sterigmatocystin, now well-known as a carcinogenic mycotoxin, was originally isolated from *Aspergillus versicolor* (3). the *o*-methyl ether is known as a metabolite of *Aspergillus flavus* (2). There is no close relationship between these latter producers and Chaetomium.

The third group of compounds consists of mollicellins A to H. They were first isolated from *Chaetomium mollicellum* and the mutagenicity of some of these was reported by American workers (16). Mollicellins G (Figure 1) and H were identified in our laboratory as products of *C. amygdalisporum* (12). These compounds belong to the depsidones, compounds commonly found in lichens.

Indole derivatives derived from tryptophan will be considered next. Biphenylbenzoquinones, called terphenylquinones, had been known as metabolites of some basidiomycetes and lichens. In cochliodinols (Figure 1), the phenyl units are replaced with indolyl groups and phenyl units are attached to the ring symmetrically. They are each assumed to be formed from two units of tryptophan and one of isoprene. Among these, cochliodinol was first isolated from *Chaetomium cochliodes* and *C. globosum* and the antibiotic effect was reported (6). We isolated the compound from *C. abuense* and *C. elatum* (20). The two positional isomers of the prenyl group were isolated from *C. murorum* and *C. amygdalisporum* and named isocochliodinol and neacochliodinol, respectively (14). These compounds exhibit some cytotoxicity to HeLa cells (IC_{50} = 5.5 μg/ml).

Chaetoglobosins (Figure 2) were isolated from *C. globosum* and the structures were elucidated in our laboratory (11). They were also isolated from *C. cochliodes, C. globosum* var. *rectum, C. mollipilium,* and *C. subglobosum,* all belonging to the *C. globosum* group (12, 20). These compounds belong to a new class of cytochalasans (17).

The phenyl group present in some cytochalasins is replaced by the indolyl group in cytochalasin A, zygosporin A and phomin (Figure 3). Biosynthesis studies using C-13 precursors revealed that they are formed from one unit of tryptophan, nine units of acetate-malonate and three C-one units (15) (Figure 4).

They exhibit a variety of effects on cultured mammalian cells. Chaetoglobosins inhibit cell division and induce poly-nucleated cells (22) (Figure 5). Fibroblastic cells grown on a solid substrate contain well-organized actin cables (also referred to as stress fibers or microfilament bundles), which can be visualized by immunofluorescence microscopy using anti-actin antibody. For instance, a C3H fibroblastic cell showed a parallel array of actin cables and a speckled distribution of actin around the nucleus. When treated with 1-20 μM chaetoglobosin A,B, D or J, a majority of the cells contracted strongly within 10 minutes. These phenomena are now known to be caused by the action of cytochalasans on the cellular protein, actin (24). Cytochalasans, including chaetoglobosins, inhibited polymerization of actin *in vitro*. (24).

The last group is epipolythiodioxopiperazines (Figure 6). Chetomin and chaetocin are known to be antibacterial and antiviral substances (18). Our studies (12, 20) revealed that chetomin is widely distrubuted among Chaetomium species. Dihydroxychaetocin, first isolated from *Verticillium tenerum,* (4) has been isolated from *Chaetomium retardatum.* (10). The causative agent of the strong cytotoxicity to HeLa cells of *C. abuense* and *C. retardatum* was recently established by X-ray analysis as a new tetrathio homolog of dihydroxychaetocin (Figure 6) and was named chetracin A (10). All these compounds exhibit strong cytotoxicity (IC_{50} about 0.03 μg/ml) and they might be among the strongest cytotoxic substances so far known.

SCREENING METHOD

The general screening method employed for our work is as follows (8). Each fungus was grown in a one-litre Roux flask on a medium composed of 200 g of polished rice and 20 ml of sterile water. The rice was previously well-washed and soaked in tap water. After inoculation, the flasks were incubated in a stationary condition at 23-25 °C for 21 days and in some cases at 30 °C for 14 days. The molded rice grains were extracted twice with dichloromethane at room temperature. The combined solution was evaporated. The residual rice grains were then extracted with ethyl acetate and the solution was evaporated. The extracts were examined by cytotoxicity tests and thin-layer chromatography. In the cytotoxicity test we used the panel method developed by Natorian Umeda (8) (Figure 7). Silica-gel F254 was used for the TLC, and mixtures of benzene-ethyl acetate and benzene-chloroform-methanol were used as the developers. Metabolites were detected by ultraviolet light usually at 254 and 365 nm. Indole derivatives, such as chaetoglobosins, were detected by the use of Ehrlich's reagent followed by heating to promote color development. Sterigmatocystin was detected by spraying with an aluminum chloride solution. For the detection of epipolythiodioxopiperazines, silver nitrate solution was employed as the spraying reagent. As far as possible, authentic samples were collected beforehand to use as standards. When a fraction showed strong cytotoxicity, attempts were made to isolate the causative agents in crystalline states for final identification. Conventional physical and chemical methods were employed for elucidating the structures of the compounds (Table 2) the deatils of which are shown in our original papers.

TABLE 2: Detection and structure elucidation of mycotoxins

TLC	Silica-gel F_{254}	
	developer: benzene-Et0Ac, benzene-$CHCl_3$-MeOH	
Detection	UV irradiation (254, 365 nm)	
	Ehrlich's reag.	- chaetoglobosins (indole derivatives)
	$AlCl_3$ soln.	- sterigmatocystins
	$AgNO_3$	- epipolythiodioxopiperazines
Structure Elucidation		
	physical methods:	UV, IR, MS, ^{1}H- and ^{13}C-, NMR, X-ray etc
	chemical reactions:	preparation of derivatives correlation reactions

THE METABOLITES OF *CHAETOMIUM VIRESCENS* (13).

Finally the separation and characterization of the metabolites of *C. virescens* will be mentioned (13). A strain of *Chaetomium thielavioideum* was found to exhibit strong cytotoxicity and the causative agents, chaetocin, chaetochromin, sterigmatocystin, and *o*-methylsterigmatocystin were isolated. The cytotoxicity and mutagenicity of these compounds, along with a related compound, are shown in Table 3.

TABLE 3: Cytotoxicity and mutagenicity of the metabolites of *Chaetomium thielavioideum*

Compound (μg/ml)	Cytotoxicity to HeLa Cells [a]							
	100	32	10	3.2	1.0	0.32	0.1	0.032
Chaetocin	4	4	4	4	4	4	3	1
Chaetochromin	4	4	4	3.5	0			
Sterigmatocystin	4	4	4	4	2	0		
O-Methylsterigmatocystin	4	4	2	1	0			
Eugenitin	0	0	0	0	0			

	Mutagenicity in *Salmonella typhimurium* [b]			
	TA 100		TA 98	
	-S-9	+S-9	-S-9	+S-9
Chaetocin	-	-	-	-
Chaetochromin	-	-	-	-
Sterigmatocystin	8	650	3	54
O-Methylsterigmatocystin	22	565	-	34
Eugenitin	-	-	-	-

a) The degree of toxicity is estimated on a scale ranging from 0 (no cellular damage) through 4 (complete cytolysis)

b) Number of revertant colonies per μg of the metabolite. Spontaneous revertant colonies were not included

The methods used to separate the metabolites are shown in Figure 8. The molded rice was extracted successively with dichloromethane and methanol. Chaetocin was precipitated owing to its low solubility. Other compounds were separated by column chromatography using silica gel. The yields of the metabolites were greatly influenced by the temperature and culture media. A good yield of metabolites was obtained by employing the solid culture on rice. Table 4 shows the differences in the relative concentrations of the metabolites following inculation at two temperatures.

TABLE 4: Summarized data on secondary metabolites of *Chaetomium thielavioideum*[a] cultured at 23° and 30°C on moldy rice

Metabolites	Yields (mg) per 100 g rice	
	30°C, 2 weeks	23°C, 3 weeks
Chaetocin	35	15
Chaetochromin	46	trace
Eugenitin	57	trace
Sterigmatocystin	4.0	0.1
O-methylsterigmatocystin	0.3	44
Ergosterol	43	trace

[a] Strain No. NHL 2827.

As a result of these studies, four mycotoxins, sterigmatocystin and *o*-methylsterigmatocystin (both carcinogenic and mutagenic compounds), chaetochromin (a teratogen), and chaetocin (an antibacterial and cytotoxic substance), were isolated and characterized as shown in Figure 8. It is noteworthy that the same mold produced different types of mycotoxins from the chemical, biogenetical and toxicological points of view.

Further work revealed that strains of *Chaetomium cellulolyticum* and *Achaetomiella virescens* also produced these same metabolites (12). This finding led to the mycological examination of these molds to establish that they are synonomyous with *Chaetomium virescens* as reported by Udagawa et al. (21).

CONCLUSION

The production of mycotoxins by Chaetomium species had been rather neglected until recently. The main reason for this neglect is assumed to be that these compounds are not acutely toxic, expecially by oral administration. Now many species of this genus are known to produce a variety of mycotoxins. It follows that considerable care should be paid to the detection of mycotoxins when *Chaetomium* spp. are employed for the production of single cell protein. Not only acute toxicity but also chronic and genetic toxicity should be carefully tested. Bioproduction of secondary metabolites differs greatly among different strains in the same species and in different cultural conditions. In order to assure the safety of microbial biomass protein, selection of the strains and examination of the cultural conditions should be carried out, with the possible formation of mycotoxins in mind.

REFERENCES

1) Brewer, D. and Taylor, A: The production of toxic metabolites by *Chaetomium* spp. isolated from soils of permanent pasture. Can. J. Microbiol., 24, 1082 (1978).

2) Burkhaardt, H.J., and Forgacs, J.: *o*-Methylsterigmatocystin, a new metabolite of *Aspergillus flavus* Link ex Fries. Tetrahedron, 24, 717 (1967).

3) Hatsuda, Y., Kuyama, S. and Terashima, N.: Studies on the metabolites of *Aspergillus versicolour*. Part 2. Nippon Nogei Kagaku Kaishi, 28, 992 (1954).

4) Hawser, D. Loosli, H.R., and Niklaus, P.: Isolierung von 11α,11'α-Dihydroxychaetocin aus *Verticillium tenerum*. Helv. Chim. Acta, 55, 2182 (1972).

5) Ito, Y. and Ohtsubo, K.: Exencephaly in mice induced by feeding chaetochromin-containing diet. Proc. Jap. Assoc. Mycotoxicol., No. 16, 22 (1982).

6) Jerram, W.A., McInnes, A.G., Maass, W.S., Smith, D.G., Taylor, A. and Walter, J.A., The Chemistry of cochliodinol, a metabolite of *Chaetomium* spp. Can. J. Chem., 53, 727 (1975).

7) Low, I., Jahn, W., Wieland, Th., Sekita, S., Yoshihira, K. and Natori, S.: Interaction between rabbit muscle actin and several chaetoglobosins or cytochalasins. Anal. Biochem., 95, 14 (1979).

8) Natori, S., and Umeda, M.: Survey of mycotoxins monitored by cytotoxicity Testing. In "Advances in Natural Products Chemistry", ed. by S. Natori, N. Ikekawa, and M. Suzuki, Kodansha, Tokyo, and Halsted Press, new York, 1981, p. 106.

9) Ohtsubo, K.: Oral toxicity of chaetochromin, a new mycotoxin produced by *Chaetomium virescens,* to mice. Proc. Jap. Assoc. Mycotoxicol., No. 12, 28 (1980).

10) Saito, T., Koyama, k., Natori, S., and Iitaka, Y.: Chetracin A, a new epipolythiodioxopiperazine having a tetrasulfide bridge from *Chaetomium abuense* and *C. retardatum,* Tetrahedron Lett., in press.

11) Sekita, S., Yoshihira, K., Natori, S. and Kuwano, H.: Structures of Chaetoglobosin A and B, cytotoxic metabolites of *Chaetomium globosum.* Tetrahedron Lett., 1973, 2109; Sekita, S., Yoshihira, K., Natori, S., Udagawa, S., Sakabe, F., Kurata, h., and Umeda, M.: Cytotoxic 10-(indol-3-yl) - (13)cytochalasans from *Chaetomium* spp. I. Production, isolation and some cytological effects of chaetoglobosins A-J. Chem. Pharm. Bull., 30, 1609 (1982).

12) Sekita, S., Yoshihira, K., Natori, S., Udagawa, S., Muroi, T., Sugiyama, Y., Kurata, H., and Umeda, M.: Mycotoxin production by *Chaetomium* spp. and related fungi. Can. J. Microbiol., 27, 766 (1981).

13) Sekita, S., Yoshihira, k., and Natori, S.: Chaetochromin, a bis(naphto-dihydropyran-4-one) mycotoxin from *Chaetomium thielavioideum:* Application of ^{13}C-^{1}H long-range coupling to the structure elucidation. Chem. Pharm. Bull., 28, 2428 (1980).

14) Sekita, S.: Isocochliodinol and neocochliodinol, bis(3-indolyl)-benzoquinones from *Chaetomium* spp. Chem. Pharm. Bull., 31, 2998 (1983).

15) Sekita, S., Yoshihira, k., and Natori, S.: Chaetoglobosins, cytotoxic 10-(indol-3-yl)-(13)cytochalasans from *Chaetomium* spp. IV. ^{13}C-Nuclear magnetic resonance spectra and their application to a biosynthetic study. Chem. Pharm. Bull.; 31, 490, (1983).

16) Stark, A.A., Kobbe, B., Matsuo, K., Buchi, G., Wogan, G., and Demain, A.L.: Mollicellins: mutagenic and antibacterial mycotoxins. Appl. Environ. Microbiol., 36, 412 (1978).

17) Tanenbaum, S.W. (Ed.), "Cytochalasins, Biochemical and Cell Biological Aspect", North holland, Amsterdam, 1978.

18) Taylor, A.: The toxicology of sporidesmins and other epipolythiadioxopiper-azines. In "Microbial Toxins - A Comprehensive Treatise". Vo. VII. ed. by S. Kadis, A. Ciegler, and S.J. Ajl, Academic Press, New York, 1971, p. 337.

19) Tertzakian, G., Haskins, R.H., Slarter, G.P., and Nesbitt, L.R.: The structure of cephalochromin. Proc. Chem. Soc., 195, (1964).

20) Udagawa, S., Muroi, T., Kurata, H., Sekita, S., Yoshihira, K., Notori, S. and Umeda, M.: The production of chaetoglobosins, sterigmatocystin, *o*-methyl-sterigmatocystin, and chaetocin by *Chaetomium* spp. and related fungi. Can. J. Microbiol., 25, 170 (1979).

21) Udagawa, S.: New or noteworth Ascomycetes from Southeast Asian soil. I. Trans. mycol. Soc. Japan 21, 17 (1980).

22) Umeda, M., Ohtsubo, K., Saito, M., Sekita, S., Yoshihira, K., Natori, S., Udagawa, S., Sakabe, F. and Kurata, H.: Cytotoxicity of new cytochalasans from *Chaetomium globosum*. Experientia, 31, 435 (1975).

23) Waksman, S.A., and Bnge, E.: Chaetomin, a new antibiotic substance produced by *Chaetomium cochliodes.* I. Formation and properties. J. Bacteriol., 48, 527, (1944).; Brewer, D., McInnes, A.G., Smith D.G., Taylor, A., Walter, J.A., Loosli, H.R. and Kis, Z.L.: Sporidesmins. Part 16. The structure of chaetomin, a toxic metabolite of *Chaetomium cochliodes,* by nitrogen-15 and carbon-13 nuclear magnetic resonance spectroscopy. J. Chem. Soc., Perkin Trans. I, 1978, p 1248.

24) Yahara, I., harada, F., Sekita, S., Yoshihira, K. and Natori, S.: Correlation between effects of 24 different cytochalasins on cellular structures and cellular events and those on actin *in vitro.* J. Cell Biol., 92, 69 (1982).

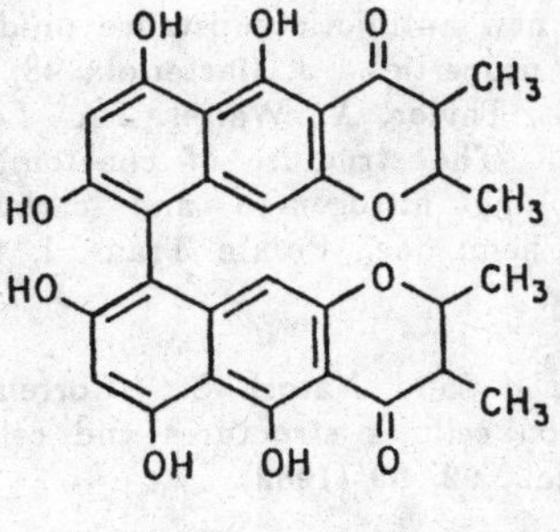

Chemical structure of chaetochromin

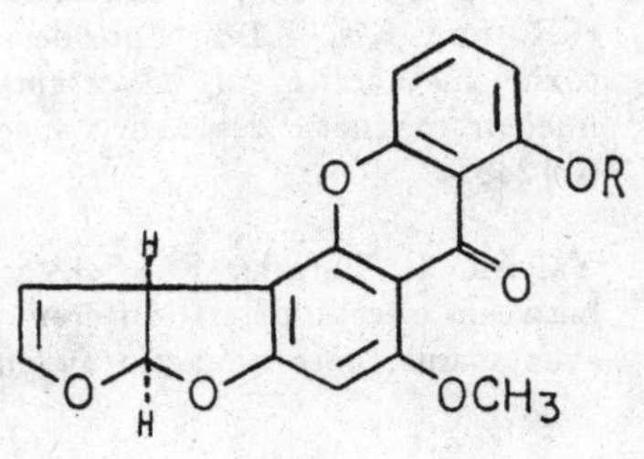

R : H STERIGMATOCYSTIN

CH_3 <u>O</u>-METHYLSTERIGMATOCYSTIN

Chemical structures of sterigmatocystin and *o*-methylsterigmatocystin

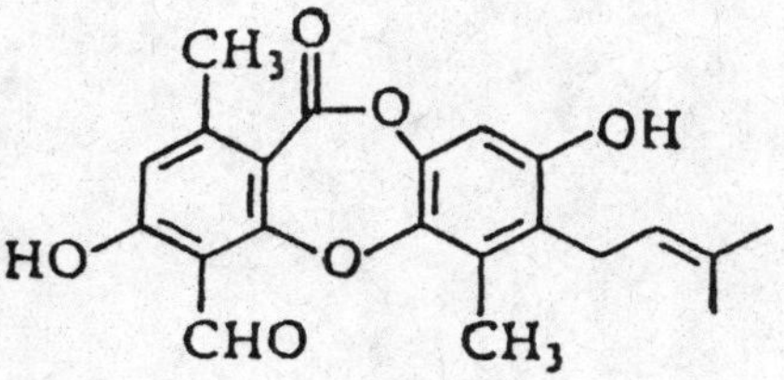

Chemical structure of Mollicellin G

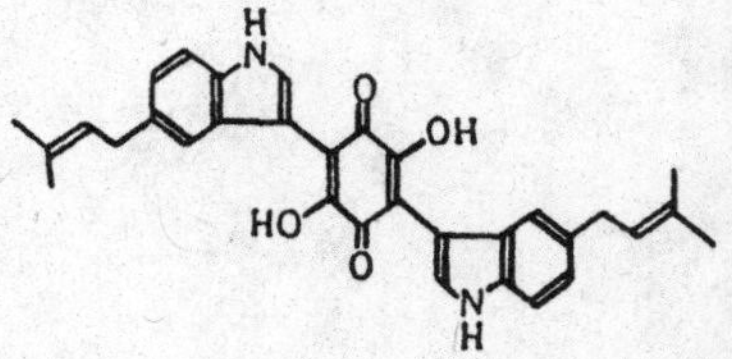

COCHLIODINOL

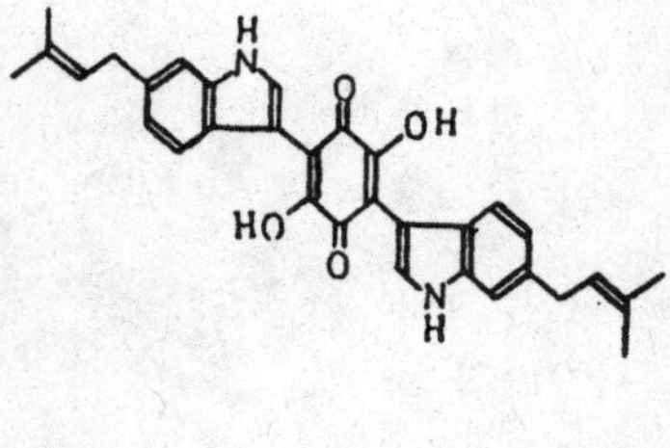

ISOCOCHLIODINOL 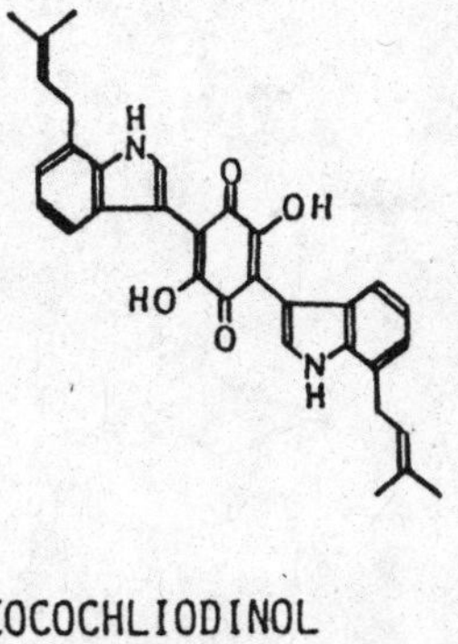 NEOCOCHLIODINOL

Chemical structures of cochliodinol derivatives

FIGURE 1: Chemical Structures

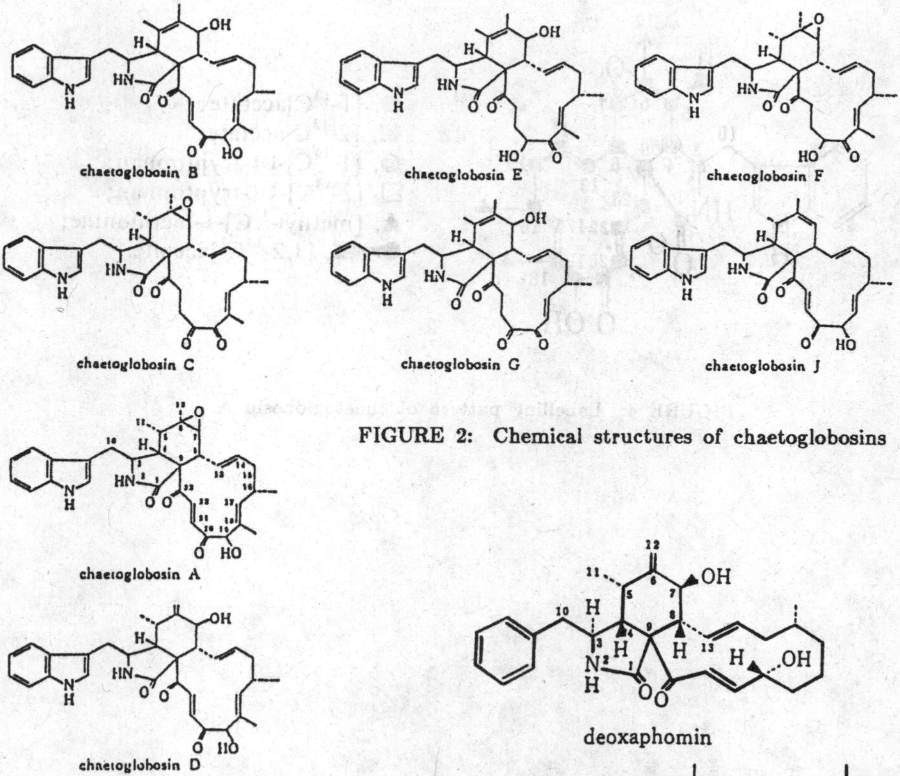

chaetoglobosin B

chaetoglobosin E

chaetoglobosin F

chaetoglobosin C

chaetoglobosin G

chaetoglobosin J

chaetoglobosin A

FIGURE 2: Chemical structures of chaetoglobosins

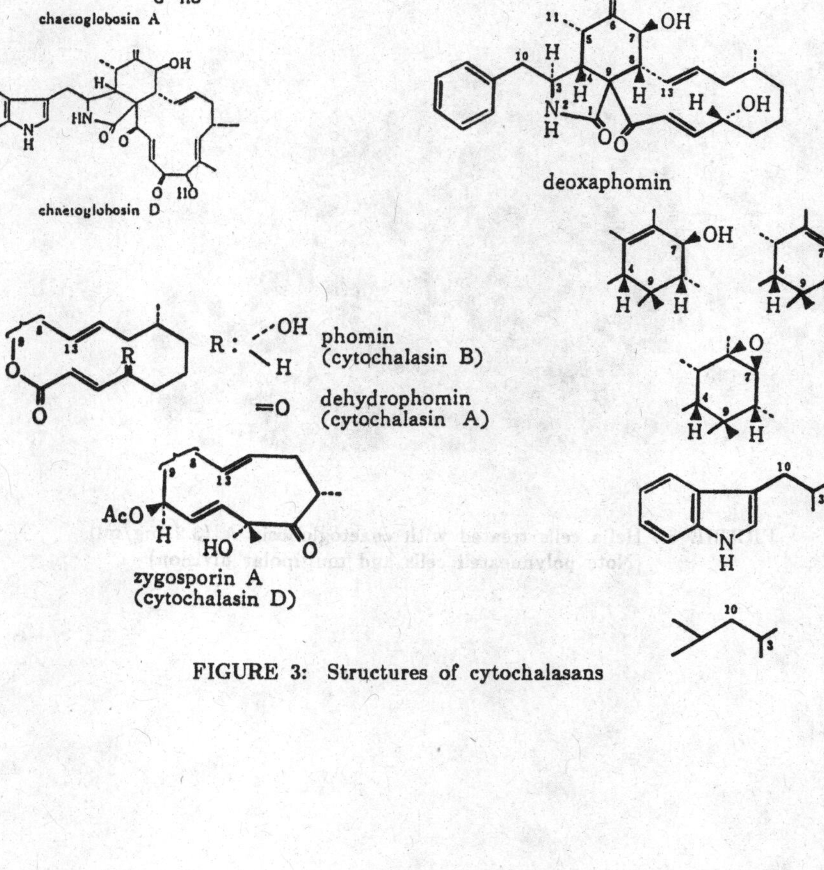

chaetoglobosin D

deoxaphomin

phomin (cytochalasin B)

dehydrophomin (cytochalasin A)

zygosporin A (cytochalasin D)

FIGURE 3: Structures of cytochalasans

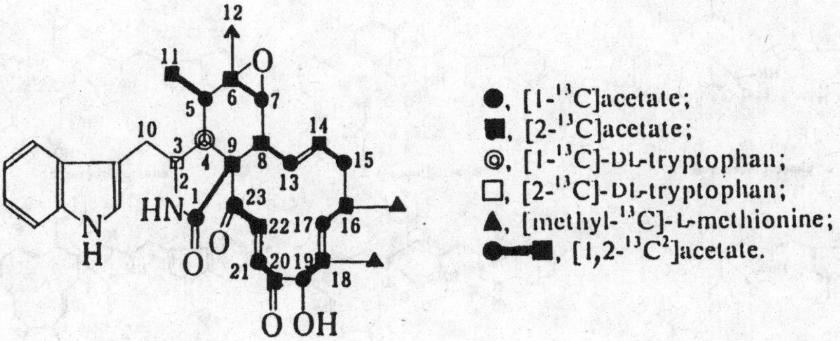

●, [1-^{13}C]acetate;
■, [2-^{13}C]acetate;
◎, [1-^{13}C]-DL-tryptophan;
□, [2-^{13}C]-DL-tryptophan;
▲, [methyl-^{13}C]-L-methionine;
●—■, [1,2-$^{13}C^2$]acetate.

FIGURE 4: Labelling pattern of chaetoglobosin A

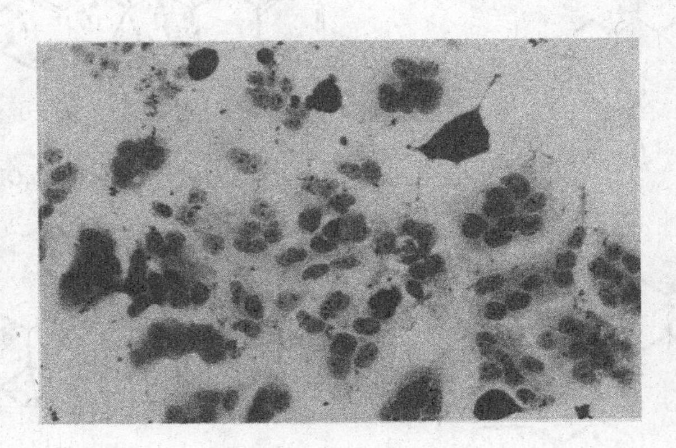

FIGURE 5: HeLa cells treated with chaetoglobosin A (3.2 mg/ml) (Note polynueated cells and multipolar division)

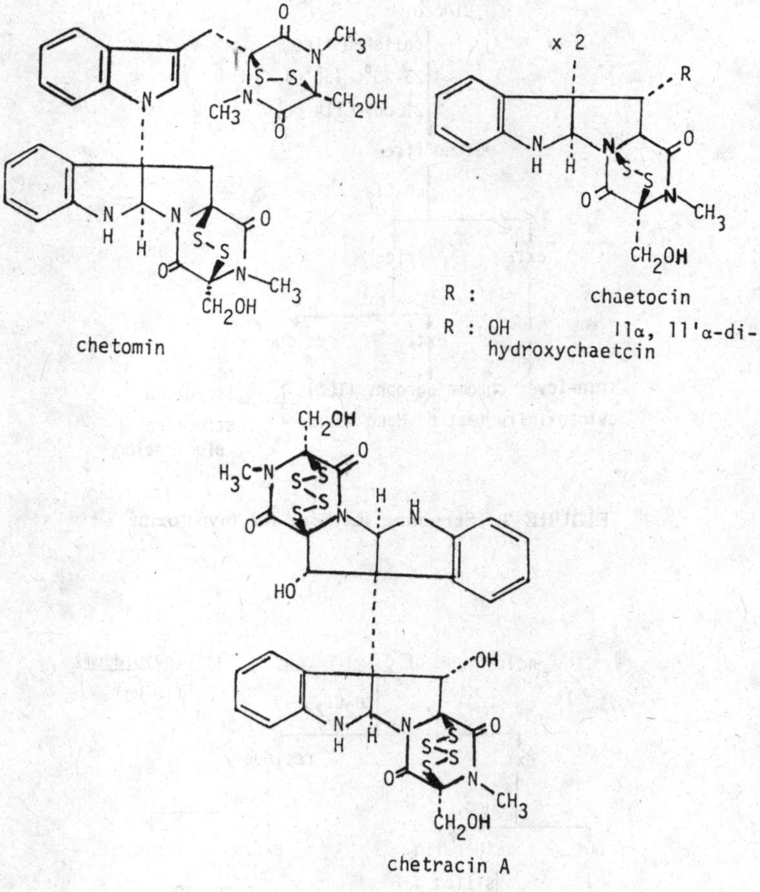

chetomin

x 2

R :

R : OH

chaetocin

11α, 11'α-dihydroxychaetcin

chetracin A

FIGURE 6: Chemical structures of epipolythiodioxopiperazines

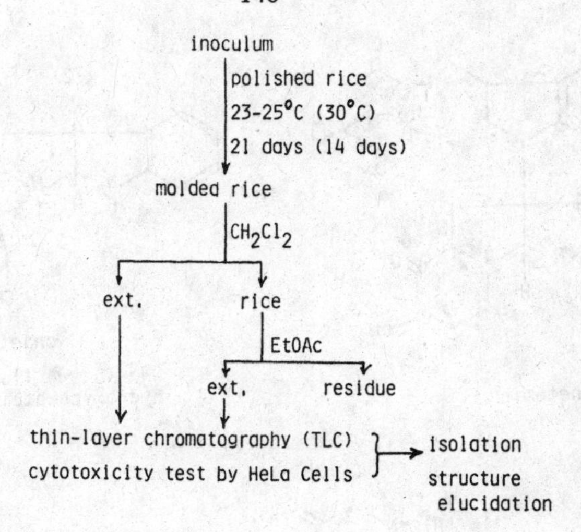

FIGURE 7: Screening methods for mycotoxins

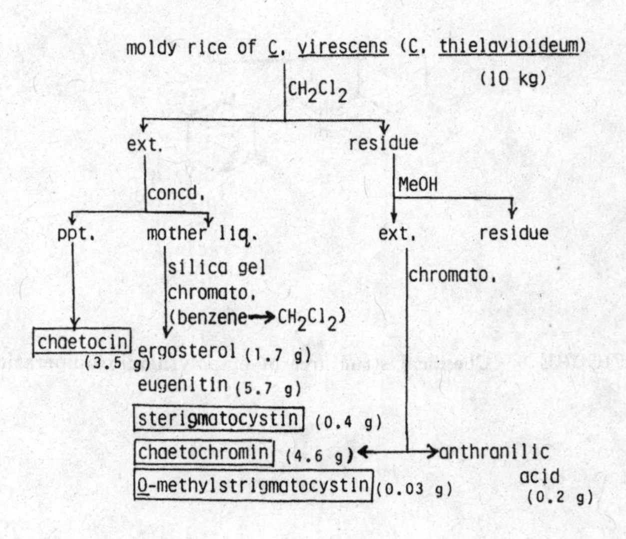

FIGURE 8: Separation of the metabolites of *C. virescens*

ALGAL BIOMASS PRODUCTION FROM WASTEWATERS AND SWINE MANURE: NUTRITIONAL AND SAFETY ASPECTS

de la Noue, J.[a,b], Proulx, D.[a], Guay, R.[a,c], Y. Pouliot[d] and J. Turcotte[a,e]

[a] Groupe de recherche en recyclage biologique et aquaculture (GREREBA), Centre de recherche en nutrition, Université Laval, Québec G1K 7P4
[b] Départment de biologie, Université Laval
[c] Départment de microbiologie, Faculté de medecine Université Laval
[d] BIONOV CNP Inc., Québec
[e] Départment de chimie, Université Laval

ABSTRACT

The biological recycling of effluents represents a means of wastewater treatment, and hence of pollution control, with the concomitant local production of animal feeds. One of the possible solutions is to grow microalgae on urban wastewaters or swine manure. The algae thus produced (Scenedesmus, Spirulina) show interesting protein content, good amino and fatty acid profiles, and can be directly used as feed for domestic animals. The algae can also be used for growing invertebrates (Daphnia), the composition of which appears appropriate for fish feeding. Contamination of the biomasses by bacteria, heavy metals or pesticides appears to be low.

INTRODUCTION

Local production of the food required by Man is probably the best solution to the problem of food shortage. The dispatch and distribution of food on a world scale raises difficulties and losses and does not permit the full utilization of overall world capacity which seems adequate, although all arable land throughout the world is in use (64).

One interesting alternative to the traditional production of food is the biological conversion of the industrial and agricultural wastes that are increasingly produced as a result of rapidly expanding urbanisation and intensive agricultural practices.

Moreover, water management is a crucial problem in developing, as well as in developed, countries. The discharge of untreated or insufficiently treated liquid wastes into the environment disturbs ecosystems which are most often especially fragile in developing countries where demographic pressure imposes strong pressure on the environment. Even in regions of developed countries, such as in Quebec, some agro-industrial practices (pig rearing, for example) constitute a real threat to water quality.

Conventional wastewater treatment systems, already in use in developed countries for decades, do not seem to be the definitive solution to pollution and eutrophication problems. The dispersal of huge quantities of the resulting inorganic nutrients (mainly ammonia, nitrate and phosphate) leads to eutrophication of the receiving bodies of water which is facilitated by the clarity of the effluents and the bioavailability of their nutrients. Moreover, this wastage is senseless since it is possible to consider these substances as nutrients for controlled biological systems for biomass production.

By combining controlled biomass production with tertiary treatment of wastewaters and similar liquid wastes it should be possible to contribute to the simultaneous solution of two major problems: food supply and protection of the natural environment (Figure 1).

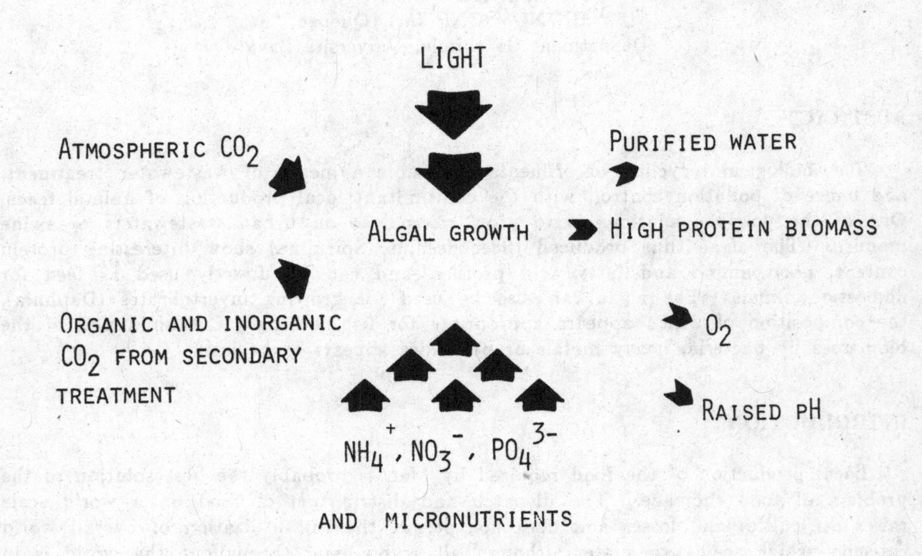

FIGURE 1: The role of microalgae in the purification of wastewater by biological recyclying

BIOMASS PRODUCTION AND BIOTREATMENTS

Traditional physicochemical tertiary treatments of wastewaters, such as ammonia stripping or phosphate precipitation, suffer from two drawbacks: expense and lack of nutrient recycling. This is not the case with biolgocial systems, from the second aspect at least. When a biotreatment is used, at least in the case of solar biotechnology (growth of microalgae on effluents), nutrient recovery can be high and leads to good biomass production linked with efficient treatment of the effluents. Thus, the environmental deterioration problem can be largely solved with little energy expense and production of biomass of good nutritional quality.

These advantages stem from the fact that microalgae use atmospheric CO_2 (or CO_2 produced by bacterial degradation of organic substrates present in wastewaters) and the macronutrients (NH^+, NO_3^-, PO_4^{3-}) present in urban and animal wastes; microalgae convert these nutrients photosynthetically into biomass with simultaneous production of oxygen, a highly useful oxidant for the completion of organic material degradation by bacteria. The use of microalgae for wastewater treatment was proposed some time ago (15,61) and many authors have proposed various systems for achieving this goal (75).

Although the production of microbial biomass proteins (MBP) from microalgae will not entirely solve the problems, it is one of the most promising means, especially when applied to wastewaters, for the direct production of food in some cases or for indirect production of food through animal feeding (46,51,85).

Energy Storage In MBP

Solar biotechnology is one of the most efficient methods of converting solar energy into food, especially when applied to algal biomass production. The maximum theoretical conversion coefficient for total light energy has been established at 6.6% (75). This percentage corresponds to the ratio between the caloric value of the autotrophic algae and the total solar energy. The actual values observed reach 2.2% (65) and compare favorably with the value of 1.5% measured for traditional agricultural production (75), although the latter could apparently reach as much as 6% (3). In practical terms, the results obtained with microalgae are comparable with the best values attained by the most efficient C4 plants, such as sugar cane and sorghum (75).

Production of MBP

The above considerations are in line with the high growth rates of algae. Algal cultures show high productivities, the daily biomass yield reaching values of 20-40 g dry matter per m^2. This corresponds to yearly average yields of 50-110 tons of dry matter (38,75) which is equivalent (dry matter basis) to 4 times the protein productivity of maize and 3 times that of soyabean (53) (see Table 1 for other data). Under favorable conditions and with intensive modes of cultivation, microalgal cultures can produce up to 20-35 times more protein than soybean, for the same area (79).

TABLE 1: **Annual protein production from various crops (36).**

Crop	Protein conc.(%)	Productivity (T/ha • y)	Annual Protein production (T/ha)[a]
Soya	16.3	13.8	2.2
Soyabean (9)	35.0	4.0	1.4
Maize (78)	7.5	29.8	2.0
Sugarcane (86)	1.8	14.0	1.0
Sugarcane (78)	-	125.3	2.3
Rice (78)	3.2	22.0	0.7
Husked rice (9)	7.1	8.0	0.6
Wheat (9)	9.5	6.7	0.6
Sorghum (9)	7.5	15.7	1.2
Spirulina (36)	61.1	12.1	7.1

a Differences in values for annual productivity for the same plant are likely to be due to the allocation of different growth periods. This period was fixed at 180 days for Spirulina in a cold temperate climate. Values are gross approximations.

Similar figures have been obtained with small-scale systems for species other than Spirulina, which indicated the potential of algal cultures. Large-scale systems will have to be developed and tested, however, before one can assess the economic competitiveness of such systems.

Various factors influence algal productivity: light, temperature and pH are the most prominent of these. The specific algal growth rate is sensitive to the first two (83,90) and the bio-availability of nutrients is substantially controlled by pH. Despite these limitations, algal cultivation is not restricted to warm sunny climates. In fact, some pilot-scale experiments carried out in Quebec (66) show that the productivity of the microalga *Scenedesmus obliquus,* cultivated on secondary effluents of urban origin (Figure 2), compares favorably with values reported by Hendricks and Bosman (43) under warmer climatic conditions. Furthermore, the experiments done in the northern cold climate of Quebec have demonstrated the feasibility of purification of wastewaters by intensive microalgal cultures throughout the year. The main problem to be solved is that of reducing the operating costs by optimization of the procedure and by the use of appropriate greenhouse technology (66).

Another aspect linked to productivity is that of water consumption. In arid or semi-arid contexts, algal cultures might be of special interest due to the lower water consumption than that required by traditional cultivars (Table 2). If one considers that the water used for algal cultures can be used afterwards for irrigation, algal cultures are even more advantageous.

FIGURE 2: Algal culture on wastewaters in a greenhouse under cold climatic conditions

TABLE 2: Comparative water consumption of various traditional cultures and Scenedesmus (44).

Culture	Water requirements (m^3) per ton of protein
Rice	69 012
Beans	32 944
Maize	29 763
Wheat	22 225
Alfalfa	6 405
Scenedesmus	863

NUTRITIONAL ASPECTS

Biomass Composition

Biomasses of the green algae Scenedesmus cultivated on artificial media, urban and piggery effluents are mainly composed of protein and carbohydrates. The protein concentration of Scenedesmus cultivated on pig slurry is superior to that of Scenedesmus grown on other liquid substrates, which in turn is superior to soyabean (Table 3).

TABLE 3: Chemical composition of *Scenedesmus sp.* powders as compared to the edible constituents of soyabean. (values in percent of dry matter)

Constituent	1	2	3	4	5	6
Crude protein	50-56	50-55	41.2	44.3	57.2	34-40
Carbohydrates	10-17	10-15	43.4	29.2	18.5	19-35
Lipids	12-14	12-19	4.1	9.5	3.0	16-20
Crude fiber	3-10	10-12	1.0	0.7	n.d.	3-5
Minerals	6-10	6-8	10.2	7.9	8.2	4-5
Humidity	4-8	5-7				7-10

1: *Scenedesmus sp.* (77). n.d.: not determined.
2: *Scenedesmus acutus* (4).
3: *Scenedesmus obliquus* urban wastewater (21).
4: *Scenedesmus obliquus* urban wastewater (68).
5: *Scenedesmus obliquus* piggery effluents (68).
6: Soyabean (77).

The differences in the ratio of protein/(sugar + lipid) concentrations observed among various sources of Scenedesmus are mainly due to the physiological status of the microalgae under various culture conditions, especially nitrogen limitation. The observed modifications of cell composition occur when the algal cultures reach the stationary phase of growth which leads to a reduction of protein concentration to the benefit of sugar and lipid concentrations (74).

Spirulina, a blue-green alga grown in Mexico (32), has a protein concentration higher than that measured for Scenedesmus or soyabean (Table 4). *Spirulina maxima,* cultivated on pig slurry, has a protein concentration similar to that of cultures grown on artificial media (Table 5).

TABLE 4: Chemical composition of Scenedesmus and Spirulina as compared to soyabean (Values in percent of dry matter)

Constituent	Scenedesmus	Spirulina[a]	Soyabean[a]
Water	4-8	10	7-10
Crude protein	50-56	50-62	34-40
Lipids	12-14	2-3	16-20
Carbohydrates	10-17	16-18	19-35
Crude fiber	3-10	-	3-5
Minerals	6-10	4-6	4-5

[a] (77).

TABLE 5: Amino acid content (g/16 g N) and total protein (% of dry weight) of *Spirulina maxima* grown on various media.

Amino acid	PS[a]	PS[b]	PS[c]	WW[d]	AM[e]	FAO[f]
ILEU	5.31	5.61	6.30	5.40	4.13	4.20
LEU	8.93	8.80	9.30	8.10	5.80	4.80
LYS	4.90	5.16	4.10	3.70	4.00	4.20
MET	2.56	4.33[g]	2.00	2.00	2.17	2.20
PHE	4.87	4.40	5.00	4.30	3.95	2.80
THR	5.36	3.84	5.20	4.50	4.17	2.80
TRY	n.m.[h]	n.m.	1.10	1.30	1.13	1.40
VAL	5.79	6.25	7.20	6.30	6.00	4.20
ALA	8.36	7.34	8.02	6.88	5.82	-
ARG	7.06	8.14	7.13	7.63	5.98	-
ASP	10.71	10.20	10.60	10.40	6.43	-
GLU	15.53	15.18	14.20	12.00	8.94	-
GLY	5.40	4.48	5.10	4.20	3.46	-
HIS	1.65	1.73	1.80	1.50	1.08	-
PRO	3.98	3.58	4.10	3.60	2.97	-
SER	4.84	4.28	4.80	4.20	4.00	-
TYR	4.78	4.86	4.10	3.70	n.m.	-
CYS	n.m.	n.m.	n.m.	n.m.	0.67	-
Total protein	61.1	75.4	63.2	60.6	71.0	-

[a] Growth on pig slurry (36).
[b] Growth on pig slurry (17).
[c] Growth on pig slurry (87).
[d] Growth on wastewaters (87).
[e] Growth on artificial medium, Durand-Chastel (29).
[f] (31).
[g] Likely to include all S-Containing amino acids.
[h] Not measured.

A comparison of the protein concentrations of various feedstuffs to those of Scenedesmus and Spirulina indicates that the latter display protein contents well above those of traditional feedstuffs, with the exception of fishmeal (18).

Amino Acid Composition of Algal MBP

When MBP's are to be used for food purposes, gross composition is not the sole characteristic to be considered and one needs, among other things, to establish the amino acid profile of the biomass. Rather than simply increasing the total protein consumption, it is necessary to provide an adequate and balanced supply of amino acids.

From Table 6, it can be seen that, for *Spirulina maxima* grown on pig slurry, the FAO amino acid requirements are easily met for all essential amino acids. It must be recalled, however, that this does not automatically lead to the conclusion the *Spirulina* sp. is nutritionally adequate: digestibility of the biomass must also be appropriate.

TABLE 6: Essential amino acid composition of *Scenedesmus sp.* compared to the FAO pattern (g/16 g N).

Amino acid	*Scenedesmus acutus*[a]	*Scenedesmus obliquus*[b]	FAO[c]
VAL	4.7	5.7	5.0
LEU	7.0	8.3	7.0
ILEU	3.1	4.1	4.0
PHE + TYR	6.0	10.1	6.0
LYS	4.6	5.9	5.5
MET + CYS	3.2	2.9	3.5
TRY	1.7	-	1.0
THR	4.9	8.6	4.0

[a] Grown on artificial medium (4).
[b] Grown on urban wastewaters (68).
[c] (31).

As shown in Table 6, the green alga *Scenedesmus obliquus* grown on urban wastewaters is relatively deficient in sulphur-containing amino acids (methionine and cysteine) compared to the FAO standards, but shows a better amino acid composition than does *Scenedesmus acutus* grown on artificial medium. It can be hypothesised that wastewaters provide micronutrients likely to be absent from artificial media or that a positive interaction occurs with the bacteria initially present in wastewaters.

Vitamins and Fatty Acid Content of Algae

The richness of microalgae in vitamins, especially those which are water-soluble, is worth mentioning (see Table 7). Microalgae are also rich in fatty acids (77,80), the most abundant being palmitic (16:0), oleic (18:1) and linolenic (18:3) acids; myristic (14:0) and linoleic (18:2) acids are less abundant. Unsaturated fatty acids account for 29-65% (dry weight basis) (48).

Algal MBP Variability in Composition and Nutritional Value

Some control can be exercised over the composition of algal biomasses in order to meet specific needs. Algae prove to be highly plastic with respect to composition according to the culture medium; nutrient levels, physico-chemical conditions and physiological state lead to different compositions for a given algal species (19, 48, 70, 84).

TABLE 7: Vitamin content (mg/100 g protein) of Scenedesmus and egg (4).

Vitamin	Egg	Scenedesmus[a]
Thiamin (B_1)	0.8	3.2
Riboflavin (B_2)	2.4	7.3
Niacinamide	0.6	13.1
Folic acid	0.04	0.15
Pantothenic	12.2	2.2
Coblamine (B_{12})	0.02	0.07
Tocopherol	7.9	26.3
Biotin	0.8	0.04
Ascorbic acid (C)	-	38.0
*B*carotene	-	45.6
Total carotenoids	-	395.0

a Probably grown on artificial medium (no details on culture medium given by the author).

TABLE 8: Protein and gross energy contents of oocystis according to processing technique used.

Protein source	Crude protein[a] (%)	True protein[b] (%)	Energy (kJ/g)
1. Algae (drum-dried)	52.1	44.1	19.5
2. Algae (deaminated)	51.6	43.4	18.9
3. Algae (irradiated)	57.7	44.7	20.6
4. Algae (autoclaved)	55.0	44.2	20.6

a As N x 6.25.
b (52).

Another important influence on composition is the processing of the algal MBP. For example, drum-drying, cooking, sun-drying or freeze-drying do not change the protein content, whereas these techniques do affect starch, sugars, lipids, vitamins and amino acids (4). Table 8 illustrates the results obtained by Lee (48) with Oocystis, a green alga similar to Scenedesmus.

The acceptability and the nutritive value of MBP are also heavily influenced by the processing technique used (4, 7, 16, 60). The drum-drying technique has the advantage of sterilizing the MBP and of improving the digestibility of MBP (4) by breaking the cell walls. This is especially important when algae are fed to non-ruminant species.

Algal MBP in human nutrition

The uses of the blue-green alga, Spirulina, for human food goes back to the Aztecs in Mexico (32). It is still consumed in the Lake Tchad region of Africa (33). Several other microalgae, especially green ones, such as Chlorella and Scenedesmus, have been studied for human consumption (58).

The various indicators used, such as protein efficiency ratio (PER), net protein utilization (NPU), biological value (BV) and digestibility coefficient (DC), show that microalgae grown on artificial media can be considered as good quality food, although the products are always inferior to casein (Table 9).

TABLE 9: Nutritional value of algae compared to casein standard (See text for abbreviations used).

Alga	PER	NPU	BV	DC
Scenedesmus[a]	1.93	65.8	80.8	81.4
Spirulina[b]	1.80	62.0	75.0	83.0
Coelastrum[b]	1.68	57.1	76.0	75.1
Casein	2.50	83.4	87.8	95.1

[a] (7,8).
[b] (63).

When one compares the biological value of Scenedesmus to that of crude soya protein, it appears that the alga is at least as good (58). Malnourished children fed microalgae have shown improvement in their conditions, their daily weight gain being around 4 g without alga and around 30 g with alga (39). With a daily intake of 5 g for children (10 g for adults), Scenedesmus proved to be physiologically well-accepted and nontoxic (39). However, it is considered that the daily intake should be below 20-35 g dry weight algae, renal disorders being likely to occur at higher intake levels owing to raised blood uric acid (5). Despite their high nutritional potential, microalgae (including Spirulina) can only supply about 5% of human energy requirements. The use of intermediate organisms as converters therefore appears desirable.

Algae for Chick and Swine Feeding

Nutritional tests with animals indicate that several algal species grown on wastewaters, and therefore not intended for human consumption, have an appropriate composition (49, 68) and show a fairly good PER in broilers. In that case, substitution of 25% of the soyabean protein with algal proteins leads to the best growth performance (56).

Feeding trials with pigs, where soyabean was replaced at a level of 30% with algae (enriched in lysine), demonstrated that the final weight of the experimental animals was as good as that of the control (88).

Several difficulties must be solved, however, if one wishes to use algal MBP in animal feeding. First, microalgae are difficult to harvest. Some promising progress has been made through the use of chitosan as an efficient, non-toxic, flocculating agent (47). Second, the relative deficiency of algal MBP in sulphur-containing amino-acids (MET and CYS) calls for supplementation in most cases. Potential toxicity at high levels also has to be taken into account. Finally, raw algal MBP does not appear to be digestible by monogastric animals and some processing is therefore required.

Use of microalgae by plankton-feeders

Probably one of the most interesting use of algal MBP's is through integrated food chains starting with liquid wastes and ending with aquatic filter-feeders. Algal removal can be efficient and higher quality biomasses can be produced, easily harvested and used for animal feeding.

Manuring ponds have been known for centuries in Asia where more or less controlled ecosystems have been used to grow MBP's and build food chains ending with shrimp or fish (2, 10, 54).

Many freshwater, salt water or hypersaline orgainsms such as Daphnia, oysters and clams or *Artemia salina* have been successfully grown on microalgae cultured on wastewaters (24, 26, 45, 68, 69, 72).

Due to its ubiquity (45) and its composition (high protein content, up to 60%) (68), the cladoceran *Daphnia magna* offers a great potential for simultaneous water purification and biomass production. Interestingly enough, daphnids occur in eutrophic environments, i.e. oxidation and stabilization ponds (1, 20, 27, 55).

By the use of an appropriate feeding regimen with algal MBP and a suitable means of harvesting (68), substantial daphnid biomasses could be collected (Table 10). The lengthening of the food chain is partly compensated for by a high energy conversion ratio of 0.4 between algae and daphnids (68, 81).

TABLE 10: Daphnid productivity from various substrates in g wet weight/L•week (68).

Substrate	Productivity	Reference
Rice bran	0.5-0.6	(25)
Fertilizer	0.01-0.06	(42)
Yeasts and fertilizer	0.1-0.28	(11)
Yeasts	0.21-0.35	(13)
Microalgae	0.5-0.7	(45)
Microalgae	0.57	(41)
Microalgae grown on pig manure	0.1	(23)
Microalgae grown on wastewaters	0.1	(26)
Microalgae grown on wastewaters	3.6	(68)

In Californian stabilization ponds, De Witt and Candland (26) reported commercial harvesting of Daphnia with yields of 1.5 tons dry weight/acre•month. It should be possible to increase this value by proper management of food supply and harvesting (69).

Upgrading the Quality of the Biomass from Algal MBP to Invertebrates

By feeding *D. magna* with *Scenedesmus obliquus*, it is possible to improve the quality of the final harvested biomass as shown in Table 11.

TABLE 11: Composition of *Daphnia magna, Scenedesmus obliquus, Artemia salina* and grower pellets for salmonids (68).

Constituent (% dry wt)	S. obliquus	D. magna	Daphina[1]	A. Salina[2]	Grower[3] pellets
Crude protein	44.3	59.5	17-60	58.0	40.0
Crude fat	9.5	9.5	4-26.7	5.1	10.0
Carbohydrate	29.2	8.7	1.1-33		-
Crude fiber	0.7	5.6	-	3.5	3.0
Ash	7.9	15.5	16-33	20.6	-
Energy content (kJ/g dry wt.)	21.2	20.0	-	-	-

[1] (45).
[2] (35).
[3] (Martin Feed Mills Ltd. (MNR-82 g).

From Table 11, it can be seen that daphnid biomasses should be *a priori* an appropriate source for trout feeding since the low level of carbohydrates in daphnids as compared to algae is favorable for rainbow trout, a poor utilizer of carbohydrates (40). It is also apparent from Table 11 that the composition of daphnid biomass can be highly variable, indicating that some control can be exercised. It is indeed known that various factors will influence the chemical composition of daphnids: stage of development (12, 45, 71) food quality (57) and availability (12, 50) and temperature (76). These changes in composition are not restricted to gross composition but apply to individual amino acids and fatty acids as well (see below).

Use of Daphnids in Aquaculture

Attractive possibilities exist for aquaculture with regards to the use of daphnid biomasses, which constitute a major part of rainbow trout diet, for example, at some periods of the year (up to 98% of gastric content) (34, 82). As shown in Table 11, the gross composition of daphnids is appropriate for salmonids.

Digestibility measurements made with rainbow trout (22) using diets incorporating 30% (dry weight basis) daphnids, have shown that the apparent digestibility coefficients measured are good, although somewhat lower than those obtained with the fish-meal reference diet (Figure 3).

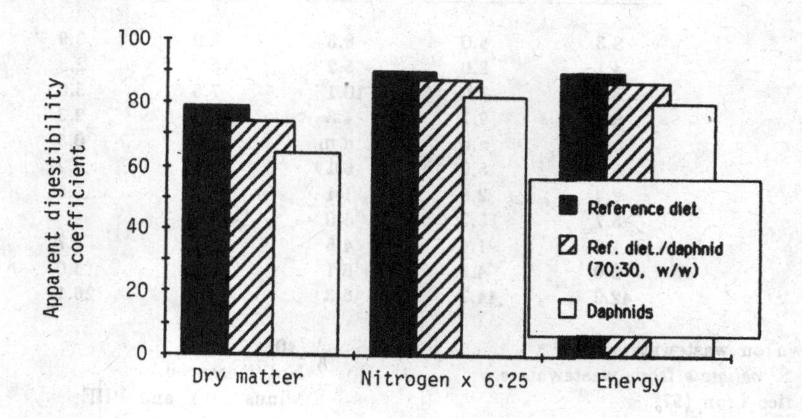

FIGURE 3: Apparent digestibility coefficients measured for the reference diet, the diet incorporating daphnid biomass and calculated for the daphnid biomass (22)

An appropriate gross composition is not sufficient *per se* to guarantee that the biomass fulfills the nutritional needs of the fish. Essential amino acids and fatty acids are also of utmost importance. Table 12 shows that the essential amino acid profile of Daphnia depends upon the food source provided and shows improvement compared to Scenedesmus and is close to that of *Artemia salina*, an invertebrate widely used in aquaculture. Moreover it appears that amino acid requirements of a Salmonid *(Onchorynchus tshawytscha*, the chinook salmon), an exacting species, should be satisfied by daphnid biomass (with the possible exception of HIS and ARG) as well as by Scenedesmus if the amino acids are bioavailable. More studies are required, however, to establish this point.

Specific nutritional needs for fish as well as for other species are not restricted to amino acids, and the fatty acid profile proves to be equally important, especially for growth and reproduction of fish. It appears, for example, that linoleic (18:2) and linolenic (18:3) acids are essential for the synthesis of long-chain polyunsaturated fatty acids in fish (40). Again, culture conditions, especially the food source, prove to be important regarding the fatty acid composition of the daphnid biomass. Table 13 illustrates this point.

TABLE 12: Essential amino acid composition of algal MBP, invertebrate biomasses and requirements of salmon (g/100 g prot.) (68).

Amino acid	*Scenedesmus*[1] *obliquus*	*Daphnia*[2] *magna*	*Daphnia*[3] *magna*	*Artemia*[4] *salina*	Amino acid[5] requirements for chinook salmon
LEU	8.3	8.0	8.3	8.0	3.9
ILE	4.1	5.0	5.2	5.3	2.5
LYS	5.9	7.0	10.1	7.6	5.0
THR	8.6	9.7	4.8	4.6	2.3
TRY	n.m.[6]	n.m.	n.m.	1.0	0.5
VAL	5.7	5.9	6.1	5.4	3.2
MET	2.4	2.8	1.1	2.7	1.5
PHE	5.7	5.7	5.0	4.7	-
HIS	1.9	1.4	4.6	1.8	1.7
ARG	6.0	4.9	6.1	6.5	6.0
Total	42.9	44.7	46.3	41.9	26.1

[1] Grown on wastewaters.
[2] Fed *S. obliquus* from wastewaters.
[3] Fed rice bran (57).
[4] (35).
[5] (40).
[6] Not measured.
[7] Minus TRY and PHE

TABLE 13: Fatty acid composition (as % of total lipid fraction) of Daphnia fed algal MBP or rice bran (68).

Fatty acid	*D. magna* fed Scenedesmus	*D. magna* fed rice bran (57)
10:0	3.8	-
12:0	0.6	-
14:0	6.9	3.9
14:1	3.4	4.1
15:0	2.2	1.0
15:1	-	0.7
16:0	25.2	21.2
16:1, w7	7.7	18.2
16:2	2.4	-
16:3	2.3	2.2
17:0	3.2	-
18:0	1.8	3.3
18:1, w9	9.7	26.0
18:2, w6	5.1	20.0
18:3, w3	17.8	0.2
20:0	trace	trace

SAFETY ASPECTS

When proposals are made for the uses of biomasses generated through biological recycling, objections can be made regarding their safety.

The first objection pertains to the nucleic acid content, usually high in MBP's. This can be accomodated by using reasonable (reduced) levels of MBP's in the diet or by using food chains (crustaceans or fish).

The second objection relates to heavy metal accumulation in the biomasses. For wastewaters, the danger appears to be minimal since the algae are grown on the secondary effluents from the treatment plant; the heavy metals therefore remain associated with the activated sludge that is decanted and discarded during the secondary treatment phase. The effluent used is therefore relatively low in heavy metals (21, 18). This is illustrated in Table 14.

TABLE 14: **Heavy metal retention of Scenedesmus biomasses produced on a domestic secondary effluent (VC) and a semi-industrial urban effluent (V) (21)**

Metal	Metal initially available on suspended matter (μg/L)		Metal initially available in solution (μg/L)		Metal in final culture medium (μg/L)		Metal in final dry biomass (μg/g)	
	VC	V	VC	V	VC	V	VC	V
Cu	11.7	20	37.1	20	36.2	20	142.3	0.2
Ni	0.5	20	6.8	20	6.5	20	10.1	1.1
Cr	0.6	20	1.9	20	1.9	20	5.7	0.2
Pb	1.0	100	4.0	100	3.1	100	24.6	1.4
Cd	-	5	-	5	-	5	-	0.03
Co	-	30	-	30	-	30	-	0.2
Fe	-	300-420	-	30-190	-	30-190	-	2.0-19.3
Mn	-	10-20	-	10-60	-	10-30	-	0.9-2.3
Zn	-	10-60	-	33-70	-	20-60	-	17-27
Hg	-	0.2-1.0	-	0.2-0.4	-	0.2-0.3	-	0.02

[a] Final biomass concentration was around 100 mg dry wt/L; pH around 9.

From data in Table 14, it appears that urban effluents may have highly different patterns of heavy metal content. Despite this variability, the concentrations measured in the secondary effluents (initial culture medium for algae) are within the limits set by Environment Canada (30) for domestic waters. Moreover, it appears that bioaccumulation of heavy metals does not occur in Scenedesmus biomasses to a significant level except in the case of copper for one of the effluents (VC, Table 14). The reason might be that algal batch cultures require 14-19 days before 50% of the heavy metals disappear from the medium (6). In our experiments, batch cultivation lasted only 7-12 days. Even in the presence of heavy metals, it has been shown that broilers fed diets that incorporated algal species grown on wastewaters and containing high levels of metals did not experience adverse effects, nor did rats fed the meat of these chickens show any detrimental effect (89). It would therefore appear that fears concerning heavy metals might be excessive, although much more detailed work is required before definite conclusions can be drawn.

The third objection to the use of MBP's obtained through biological recycling of wastewaters pertains to the possible accumulation of pesticides and organochlorinated compounds. Table 15 shows that there is essentially no such compound accumulated in algal MBP biomasses.

TABLE 15: Organochlorinated compound retention of Scenedesmus biomasses produced on a domestic secondary effluent (VC) and a semi-industrial urban effluent (V) (21).

Organochlorinated compound	Initial concentration in effluent (μg/L)		Final concentration in effluent (μg/L)		Concentration in final dry biomass (μg/g)	
	VC	V	VC	V	VC	V
Heptachlor	2	4	2	4	0.1	0.1
DDE[b]	2	6	2	6	0.1	0.1
Dieldrin	2	8	2	8	0.1	0.1
Aldrin	2	4-8	2	4-8	0.1	0.1
PCB[c]	20	30	20	30	5	2-5

[a] Same conditions as Table 14.
[b] DDT derivitive
[c] Polychlorobiphenyls.

The last concern relates to pathogens. Coliforms and pathogenic bacteria are reduced by 50 and 90% respectively through primary and secondary treatments (14). Further reduction or elimination is assured by chlorination (73). Table 16 reports some of the results we have obtained regarding pathogenicity of the biomasses produced on either wastewaters or swine manure.

TABLE 16: Total bacterial counts of biomasses obtained through biological recycling of wastewaters and swine manure (1 g dry weight samples).

Biomass	Blood agar		MacConkey agar	
	22 °C	37 °C	22 °C	37 °C
Scenedesmus[a]	27	50	0	0
Scenedesmus[b]	110	103	0	0
Daphnids[a]	0	12	0	0
Daphnids[b]	0	0	0	0

[a] Grown on wastewaters.
[b] Grown on swine manure.

Although not all identified, almost all the observed colonies were typical *Bacillus* spp. and, on a few occasions, of the Gram negative rod-type (likely Pseudomonas). No Salmonella or Shigella appeared. From this, we may conclude that algal MBP's or daphnid biomasses produced through biological recycling are probably safe regarding pathogens. In any event, remaining pathogens and bacteria would be eliminated at the processing stage of the biomasses (see above).

Work carried out in our laboratory has shown that the bacterial population drops during the exponential phase of algal culture (36). This might be due to unfavourable conditions for bacteria caused by algal growth (raised pH, etc.). Another possible explanation, to be tested, is that bacteriostatic substances could be excreted by algae when the population of the latter increases. Conversely, it would appear that the presence of bacteria is beneficial to the algae at the start of culture. Bacteria have been shown to interact positively with daphnids (67). More work is needed to explore these complex interactions.

CONCLUSION

When one considers the problem of food production for Man, it appears that the solution cannot be unique and that technical aspects are not the only ones to be considered. Economical and socio-political factors are also at play in this complex matter. Exploring new options is a necessity, and biological recycling is probably one of the possibilities to be evaluated more intensively.

Algal production from wastewaters and animal slurries is not new and many efforts have been made since the forties to design and operate systems that can prove functional and economical. Economics are still the bottleneck for commercial operation of such systems. Some changes in group (and individual) psychology will have to be made before societies recognize the economic impact of water depollution. Although figures vary according to the authors, it appears that a sizable net income can be generated through algal culture on wastewaters (28).

The biotechnological era officially started quite recently and one can expect that MBP production, especially through solar biotechnology, will be increasingly considered as a valuable contribution to biotreatments of effluents. Disposal of wastes represents a costly solution to pollution problems. In 1970, the US produced 70 x 10^5 metric tons of expensive fertilizers, but, at the same time 8-13 x 10^5 metric tons of nitrogen were lost through wastewater discharge (37). Biological recycling could, in theory, provide some 50-80 x 10^6 metric tons of MBP. This could represent a sizeable food production that cannot be ignored on entirely prejudicial grounds.

ACKNOWLEDGMENTS

The authors wish to acknowledge the financial assistance of various granting organizations (Natural Sciences and Engineering Research Council of Canada, Fonds FCAC team grant, Conseil des recherches et services agricoles du Québec and Agriculture Canada). The authors wish to thank A. Lavoie, T. Pouliot, C. Cloutier, Bl Ni Eidhin, L. Barbeau and G. Gagnon for their help during the experimental work and the preparation of the manuscript.

REFERENCES

1) D.G. Alexander, K.J. Supeene, B.C. Chu and H.D. Maciorowski. The lethal and sublethal effects of secondary-treated sewage effluent on various fish and invertebrates. Fish. Res. Board Can., Tech. Rep., no 709 (1977).

2) J.E. Bardach, J.H. Ryther and W.O. McLarney. Aquaculture - The Farming and Husbandry of Freshwater and Marine Organisms. WileyInterscience, New York (1972).

3) J.A. Bassham. Increasing crop production through more controlled photosynthesis. Science, 197, 630-638 (1977).

4) E.W. Becker. Major results of the Indo-German algal project. Arch. Hydrobiol. Beih./Ergeb. Limnol., 11, 23-40 (1978).

5) E.W. Becker. The legislative background for utilization of microalgae and other types of single cell protein. Arch. Hydrobiol. Beih/Ergeb. Limnol., 11, 56-64 (1978).

6) E.W. Becker. Limitations of heavy metals removal from wastewater by means of algae. Water Res., 17, 459-466 (1983).

7) E.W. Becker, L.V. Venkataraman and P.M. Khanum. Effect of different methods of processing on the protein efficiency ratio of the green alga *Scenedesmus acutus*. Nutr. Rep. Int., 14, 305-314 (1976).

8) E.W. Becker, L.V. Venkataraman and P.M. Khanum. Digestibility coefficient and biological value of the proteins of the alga *Scenedesmus acutus* processed by different methods. Nutr. Rep. Int., 14, 457-466 (1976).

9) J. Berend, E. Simovitch and A. Ollian. Economic aspects of algal animal food production. In Algae Biomass, Production and Use, G. Shelef and C.J. Soeder (Eds), Elsevier/North Holland Biomedical Press, Amsterdam, 799-818 (1980).

10) R. Billard. La pisciculture en etang. Comptes-Rendus du Colloque sur la Pisciculture en Etang, R. Billard (Ed), Arbonne-la-Foret, INRA Publs, Paris, 269-281 (1980).

11) I.B. Bogatova and M.K. Askerov. Experience in large-scale breeding of water flea *Daphnia magna*. Biol. Abst. No 22742 (Rybn. Khoz., 12, 21-26, 1958), (1965).

12) I.B. Bogatova, M.A. Shcherbina, V.V. Ovinnikova and N.A. Tagirova. The chemical composition of certain planktonic animals under different growing conditions. Hydrobiol., 7, 39-43 (1973).

13) M.M. Briskina. Methods for the production of life food for young fish. In Materialy Soveshchaniya Povoprosam Rybovodstva, Sbornik (Ed) (After I.V. Ivleva, 1973, Moskva) (1960).

14) F.L. Bryan. Diseases transmitted by foods contaminated by wastewater. In Environ. Protect. Agency, Proceedings of Conference on: Wastewater Use in the Production of Food and Fiber. Ser. EPA 660/2-74-041, USA, 16-45 (1974).

15) D.H. Caldwell. Sewage oxidation ponds - Performance, operation and design. Sewage Works J., 18, 433-458 (1946).

16) P.R. Cheeke, E. Gasper, L. Boersma and J.E. Oldfield. Nutritional evaluation with rats of algae (Chlorella) grown on swine manure. Nutr. Rep. Intern., 16, 579-585 (1977).

17) R.J. Chiu, H.I. Liu, C.C. Chen, Y.C. Chi, H. Sha, P. Soong and P.L.C. Hao. The cultivation of *Spirulina platensis* on fermented swine manure. In Animal Wastes Treatment and Utilization, Proc. int. Symp. Biogas, Microalgae and Livestock, Chung Po (Ed), Taiwan, 435-446 (1980).

18) G. Clement, H. Durand-Chastel and V. Henny. Une nouvelle algue alimentaire. Overdruk uit Voeding, 30, 772-781 (1969).

19) D.M. Collyer and G.E. Fogg. Studies on fat accumulation by algae. J. Exp. Bot., 6, 256-275 (1955).

20) G.R. Daborn, J.A. Hayward and T.E. Quinney. Studies on *Daphnia pulex* Leydig in sewage oxidation ponds. Can. J. Zool., 56, 1392-1401 (1978).

21) J. de la Noue, C. Thellen and R. Van Coille. Traitements tertiaires d'eaux usées municipales par production de biomasse d'algues. Rapport de recherche Agriculture Canada (1983).

22) J. de la Noue et G. Choubert. Apparent digestibility of invertebrates biomasses by rainbow trout. Aquaculture, 50, 103-112 (1985).

23) N. De Pauw, L. De Leeheer Jr., P. Laureys, J. Morales and J. Reartes. Cultures d'algues et d'invertebres sur déchets agricoles. In la Pisculture en Etang, R. Billard (Ed), INRA, Paris, 189-214 (1980).

24) N. De Pauw, H. Verlet and L. De Leeheer. Heated and unheated outdoor cultures of marine algae with animal manure. In Algae Biomass, Produciton and Use, G. Shelef and C.J. Soeder (Eds), Elsevier/North-Holland Biomedical Press, Amsterdam, 315-341 (1980).

25) N. De Pauw, P. Laureys and j. Morales. Mass cultivation of *Daphnia magna* (Straus) on ricebran. Aquaculture, 25, 141-152 (1981).

26) J.W. De Witt and W. Candland. The water flea. The American Fishfarmer, January, 8-10 (1971).

27) R. Dinges. The availability of Daphnia for water quality improvement and as an animal food source. In Environmental Protection Agency, Proceedings Wastewater Use in the Production of Food and Fiber. APE-660/2-74-041, 142-161 (1974).

28) Z. Dubinsky, S. Aaronson and T. Berner. Some economic considerations in the mass culture of microalgae. In Algae Biomass, Production and Use, G. Shelef and C.J. Soeder (Eds), Elsevier/North-Holland Biomedical Press, Amsterdam, 51-64 (1980).

29) H. Durand-Chastel. Production and use of Spirulina in Mexico. In algae Biomass, Production and Use, G. Shelef and C.J. Soeder (Eds), Elsevier/North-Holland Biomedical Press, Amsterdam, 51-64 (1980).

30) Environnement Canada. References sur la qualité des eaux. Rapport Direction generale de la qualité des eaux, Environnement Canada, Ottawa (1980).

31) FAO/WHO. Energy and protein requirement. Food and Agriculture Organization of United States, Food Nutrition Meeting Report, Series no 53, Rome (1973).

32) W.V. Farrar. Tecuitlatl; a glimpse of Aztec food technology. Nature, 211, 341-342 (1966).

33) O. Faucher, B. Coupal and A. LeDuy. Utilization of seawater-urea as a culture medium for *Spirulina maxima*. Can. J. Microbiol., 25, 752-759 (1979).

34) M.G. Galbraith Jr. Size-selective predation on Daphnia by rainbow trout and yellow perch. Trans. Am. Fish. Soc., 96, 1-10 (1967).

35) M. Gallagher and W.D. Brown. Composition of San Francisco Bay brine shrimp (*Artemia salina*). J. Agric. Food Chem., 23, 630-632 (1975).

36) J. Gauthier, P. Talbot, G. Lessard, Y. Pouliot and J. de la Noue. Epuration du lisier de porc et production de spiruline. Rapport de recherche, Compagnie Lallemand et Université Laval (1985).

37) J.C. Goldman and J.H. Ryther. Waste reclamation in an integrated food chain system. In Biological Control of Water Pollution, J. Tourbier and R.W. Pierson Jr. (Eds), University of Pennsylvania Press, Philadelphia, 197-214 (1976).

38) J.C. Goldman. Outdoor algal mass cultures. I. Applications. Water Res., 13, 1-19 (1979).

39) R. Gross, U. Gross, A. Ramirez, K. Cuadra, C. Collazos and W. Feldheim. Nutritional tests with green alga Scenedesmus with healthy and malnourished persons. Arch. Hydrobiol. Beih./Ergeb. Limnol., 11, 161-173 (1978).

40) J.E. Halver. Fish Nutrition. Academic Press, New York (1972).

41) D.G. Heimbuch. A flow-through system for the mass cultivation of Daphnia production potentials and effects on water quality. Master's thesis, Cornell University, USA (1978).

42) G. Heisig. Mass cultivation of *Daphnia pulex* in ponds: the effect of fertilization, aeration and harvest on the population development. In Cultivation of Fish Fry and its Life Food, Styczynska-Jurewicz et al. (Eds), European Mariculture Society, Special Publication, 4, 335-359 (1979).

43) F. Hendricks and J. Bosman. The removal of nitrogen from an inorganic industrial effluent by means of intensive algal culture. Prog. Wat. Tech., 12, 651-665 (1980).

44) P. Heussler, J.S. Castillo and F. Merino. Ecological balance of algal cultures in arid climates: major results of the Peruvian-German microalgae project at Trujillo. Arch. Hydrobiol. Beih./Ergeb. Limnol., 11, 17-22 (1978).

45) I.V. Ivleva. Mass Cultivation of Invertebrates: Biology and Methods. Israel Program for Scientific Translations, Jerusalem (1973).

46) R. Kihlberg. The microbe as a source of food. Ann. Rev. Microbiol., 26, 427-466 (1972).

47) A. Lavoie and J. de la Noue. Harvesting microalgae with chitosan. J. World Maricul. Soc., 14, 685-694 (1983).

48) B.H. Lee. Production et évolution de protéines d'algues unicellulaires. Ph.D. Thesis, Laval University, Canada (1980).

49) B.H. LEe, G. Picard and G. Goulet. Effects of processing methods on the nutritive value and digestibility of Oocystis alga in rats. Nutr. Rep. Intern., 25, 417-429 (1982).

50) V.H.W. Lemcke and W. Lampert. Veruanderungen im gewicht und der chemischen zusammensetzung von *Daphnia pules* im hunger (Changes in weight and chemical composition of *Daphnia pulex* during starvation). Arch. Hydrobiol., Suppl., 48, 108-137 (1975).

51) J. Lovland, J.M. Harpers and E.L. Firey. Single Cell protein for human food. A review. Lebensm.-Wiss. u. Technol., 9, 1310142 (1976).

52) O.H. Lowry, N.J. Rosebrough, A.L. Farr and R.J. Randall. Protein measurement with the folic phenol reagent. J. Biol. Chem., 193, 265-275 (1951).

53) R.H.T. Matoni, E.C. Keller and H.N. Myrick. Industrial photosynthesis, a means to a beginning. BioScience, 15, 403-407 (1965).

54) Ch. Meske and G. Pfeffer. Growth experiments with carp and grass carp. Arch. Hydrobiol. Beih./Ergeb. Limnol., 11, 98-107 (1978).

55) B.D. Mitchell and W.D. Williams. Population dynamics and production of *Daphnia carinata* (King) and *Simocephalus exspinosus* (Koch) in waste stabilization ponds. Aust. J. Mar. Freshw. Res., 33, 837-864 (1982).

56) S. Mokady, S. Yannai, P. Einav and Z. Berk. Nutritional evaluation of the protein of several algae species for broilers. Arch. Hydrobiol. Beih./Ergeb. Limnol., 11, 89-97 (1978).

57) J. Morales. Culture en masse de *Daphnia magna* Straus (Crustacea, cladocera) sur dechets bio-industriels. D.Sc. Thesis, Gand University, Belgium (1983).

58) H. Muller-Wecker and E. Kofranyi. Zur bestimmung der biologischen wertigkeit von nahrungsproteinen. 18. Mitt. einzeller als zusautzliche nahrungsquelle. Hoppe Seiler's Z. Physol. Chem., 354, 1034-1042 (1973).

59) B. Myrand and J. de la Noue. Croissance individuelle et dynamique de population de *Daphnia magna* en cultures dans les eaux usées traitées. Hydrobiologia, 97, 167-177 (1982).

60) P.T. Omstedt and A. Von der Decken. Effect of processing on the nutritive value of *Saccharomyces cerevisiae*, *Scenedesmus obliquus* and *Spirulina platensis* measured by protein synthesis *in vitro* in rat skeletal muscle. In Single Cell Protein II, S.R. Tannenbaum and D.I.C. Wang (Eds), M.I.T. Press, Cambridge, mass, 553 (1975).

61) W.J. Oswald and H.B. Gotaas. Photosynthesis in sewage treatment. Trans. Am. Soc. Civil Eng., 122, 73-105 (1957).

62) W.J. Oswald and C.G. Golueke. Harvesting and processing of waste grown microalgae. In Algae, Man and the Environment, D.F. Jackson (Ed), Syracuse University Press, Syracuse, NY, 371-389 (1968).

63) W. Pabst. Ernahrungsversuche zur bestimmung der proteinqualitat von mikroalgen. I. Symp. Mikrobielle Proteingewinnung. Verlag Chemie. (Weinheim), 173-178 (1975).

64) D. Pimentel, W. Dritschilo, J. Krummel and J. Kutzman. Energy and land constraints in food protein production. Science, 190, 754-761 (1975).

65) Y. Pouliot and J. de la Noue. Utilisation des microalgues pour le traitement tertiaire des eaux usées. Comptes-rendus, 7e symposium sur le traitement des eaux usées, Montréal, novembre (1984).

66) Y. Pouliot and J. de la Noue. Système biotechnologique solaire combiné pour la production de biomasses d'algues et l'épuration des eaux usées: aspects technologiques. Abst. Biennal Congress of the International Solar Energy Society, E. Birgen and K.G.T. Hollands (Eds), Montreal, juin (1985).

67) D. Proulx, R. Lesel and J. de la Noue. Growth of *Daphnia magna* in axenic, monoxenic and holoxenic conditions. Rev. Fr. Sci. Eau, 3, 83-91 (9184).

68) D. Proulx and J. de la Noue. Growth of *Daphnia magna* on urban wastewaters tertiarily treated with *Scenedesmus* sp. Aquacult. Eng., 4, 93-111 (1985).

69) D. Proulx and J. de la Noue. Harvesting *Daphnia magna* grown on urban tertiary-treated effluents. Water Res., 19, 1319-1324 (1985).

70) B. Richardson, D.M. Orcutt, H.A. Schuertner, C.L. Martinez and H.E. Wickline. Effects of nitrogen limitation on the growth and composition of unicellular algae in continuous culture. Appl. Microbiol., 18, 245-250 (1969).

71) S. Richman. The transformation of energy of *Daphnia pulex*. Ecol. Monog., 28, 273-291 (1958).

72) J.H. Ryther, W..M. Dunstan, K.R. Tenore and J.E. Huguenin. controlled eutrophication - Increasing food production from the sea by recycling human wastes. BioScience, 22, 144-152 (1972).

73) J.H. Ryther and K.R. Tenore. Integrated systems of mollusk culture. In Harvesting Polluted Waters Waste Heat and Nutrient-Loaded Effluents in the Aquaculture, O. Devik (Ed), Plenum press, New York, 153-167 (1976).

74) R. Samson. Caractéristiques physiologiques de la croissance et de la production de polysaccharides chez neuf éspèces d'algues endogènes isolées des effluents du traitement secondaire de l'usine d'épuration de Valcartier. Master Thesis, Laval University, Canada (1980).

75) G. Shelef and C.J. Soeder (Eds). Algae Biomass-Production and Use. Elsevier/North-Holland Biomedical Press, Amsterdam (1980).

76) N.B. Snow. The effect of season and animal size on the calorific content of *Daphnia pulicaria* (Forbes). Limnol. Oceanogr., 17, 909-913 (1972).

77) C.J. Soeder, H. Muller-Wecker, W. Pabst and H. Grant. Ann. Hyg. L. Fre.-Med. et Nutr., 6, 49 (1970).

78) C.J. Soeder. Primary production of biomass in freshwater with respect to microbial energy conversion. In Microbial Energy Conversion, H.G. Schlegel and J. Barnea (Eds), Erik Goltzke KG, 59-68 (1976).

79) L. Switzer. Spriulina, the Whole Food Revolution. Proteus Corporation, Bantam Books (1982).

80) Y. Takechi. Chlorella: Íts Fundamental and Application. Gakken Co., Tokyo, Japan (1971).

81) E. Tarifeno-Silva, L. Dawasake, D.P. Yu, M.S. Gordon and D.J. Chapman. Aquacultural approaches to recycling of dissolved nutrients in secondarily treated domestic wastewaters - II. Biological productivity of artificial food chains. Water Res., 16, 51-57 (1982).

82) W.W. Taylor and S.D. Gerking. Population dynamics of *Daphnia pulex* and utilization by the rainbow trout (*Salmo gairdneri*). Hydrobiologia, 71, 277-287 (1980).

83) L. Van Liere. On *Oscillatoria agardhii* gomont: Experimental ecology and physiology of nuisance bloom-forming cyanobacterium. Ph.D. Thesis, Amsterdam University (1979).

84) L.V. Venkataraman, E.W. Becker and P.M. Khanum. Supplementary value of the proteins of algae *Scenedesmus acutus* to rice, ragi, wheat and peanut proteins. Nutr. rep. internat., 15, 145-155 (1977).

85) C.I. Waslien. Unusual sources of proteins for man. Crit. Rev. Food Technol., June, 77-151 (1975).

86) O.W. Wilcox. Footnote to freedom from want. J. Agr. Food Chem., 7, 12 (1959).

87) J.F. Wu and W.G. Pond. Amino acid composition and microbial contamination of *Spirulina maxima,* a blue-green alga, grown on the effluent of different fermented animal wastes. Bull. Environm. Contam. Toxicol., 27, 151-159 (1981).

88) L.B. Yang, L.K. Wing, M.G. McGarry and M. Graham. Wastewater Treatment and Resource Recovery. IDRC Report 154^{e}, International Development Research Center, Canada (1980).

89) S. Yannai, S. Mokady, K. Sachs and Z. Berk. The safety of several algae grown on wastewater as a feedstuff for broilers. Arch. Hydrobiol. Beih./Ergeb. Limnol., 11, 139-149 (1978).

90) W. Zevenboom. Growth and nutrient kinetics of *Oscillatoria agardhii.* Ph.D. Thesis, Amsterdam University (1980).

NUTRITIVE VALUE OF METHANE FERMENTATION RESIDUE PRODUCED FROM CATTLE AND SWINE WASTES

D.N. Mowat, C.R. Jones, J.G. Buchanan-Smith and G.K. Macleod

Department of Animal and Poultry Science
University of Guelph
Guelph, Ontario, Canada
N1G 2W1

ABSTRACT

Methane fermentation residue (MFR) produced commercially from cattle or swine wastes was evaluated as a protein supplement for growing beef cattle. The suspended solids, removed from the fermentation effluent by centrifugation, contained 24.7 and 37.5% crude protein in the dry matter, respectively. Steers fed traditional urea or soybean meal supplements had markedly greater weight gains and feed efficiencies compared with steers supplemented with cattle MFR. The reduced performance was due in part to poor nutrient digestibilities of the MFR. It was concluded that MFR produced from cattle wastes has little, if any, feeding value for beef cattle. While swine MFR appeared to have greater potential as a feed source due to its higher nitrogen content and improved nutrient digestibilities, in a subsequent steer growth trial, this material also markedly depressed weight gain and feed efficiency. A mineral imbalance may be involved.

INTRODUCTION

Methane fermentation of animal wastes has received considerable attention since the energy crisis of the mid 1970's. The process not only produces energy for potential use on a farm but also reduces odors in the liquid fertilizer. Operational units currently exist on a few Canadian (and U.S.) farms. However, for most locations in Canada, methane production alone is not sufficient justification at least with current energy prices.

Recently, interest developed in the production and potential feeding of MFR which could alter economics of the overall process. Fermented effluent from the mesophilic anaerobic process is centrifuged to produce the semi-solid residue and a liquid fertilizer (Macdonald et al., 1984). This residue was supposedly high in crude protein and mineral content and deemed to have potential as a supplement to diets particularly for beef cattle.

This research was conducted to investigate the nutritive value of MFR produced from a commercial beef feedlot operation and a swine enterprise as protein supplements for growing beef cattle.

MATERIALS AND METHODS

Chemical composition

Samples of MFR were collected from two fermentors utilizing livestock wastes located on large commercial farms in Ontario. One farm fattened cattle in a slatted-floor facility on a diet of corn grain and silage (75%) and byproducts from food processing (25%). The other farm grew and finished pigs on a corn grain and soybean meal diet. Both fermentation plants were designed and operated under similar conditions (Macdonald et al., 1984). The suspended solids were removed from the fermentation effluent by centrifugation. A total of 12 samples were obtained from each fermentor at bi-weekly intervals over a six-month period between spring and autumn 1983. Samples were analyzed for numerous constituents (Buchanan-Smith et al., 1984).

Feedlot trial 1

Eighty beef steer calves averaging 296 kg liveweight were fed high corn silage diets as outlined in Table 1 for a period of 84 days. One diet containing cattle MFR (MFR_1) was formulated to be isonitrogenous and isocaloric with the urea or soybean meal control diets. The second diet containing cattle MFR (MFR_2) was formulated to be isonitrogenous but with equal ratios of corn silage to high moisture corn as the soybean meal diet, resulting in a lower estimated available energy content. Cattle were fed once daily *ad libitum*. Further procedural details have been published (Jones et al., 1986).

Digestion trial

Six steers averaging 410 kg were used in a 3 x 3 double latin square design to determine digestibilities of cattle and swine MFR. Cattle or swine MFR replaced 13% of a diet similar to the soybean meal diet used in feedlot trial 1. Routine procedures were followed (Jones et al., 1986).

Feedlot trial 2

Forty large-framed steer calves averaging 263 kg liveweight were used to determine performance of cattle fed swine MFR. A corn silage and high moisture shelled corn basal diet was supplemented with either soybean meal and urea (control) or swine MFR at a level of 10% of diet dry matter for the first 56 days and 7.6% for days 57-112.

TABLE 1: Composition of diets fed during feedlot trial 1

Items	Nitrogenous feed source				
	Urea	SBM	MFR_1	MFR_2	SEM+
Ingredients (% DM)					
Corn silage	88.8	89.3	61.0	81.0	
High moisture corn	7.5	3.0	22.5	2.7	
Urea	0.6	-	-	-	
Soybean meal	-	4.6	-	-	
MFR cattle	-	-	14.0	13.8	
Mineral 1*	1.4	1.4	-	-	
Mineral 2**	-	-	0.7	0.7	
Rumensin premix***	1.8	1.8	1.8	1.8	
Chemical Composition					
Dry matter (%)	43.3^a	43.4^a	38.4^b	35.7^c	0.29
Crude protein (%DM)	10.9^a	11.4^b	11.9^c	11.6^b	0.09
TDN estimated (%DM)++	71	71	71	66	
ADF (%DM)	21.3^a	21.5^a	19.6^b	24.0^c	0.23
NDF (%DM)	36.0^a	36.3^a	31.1^b	37.6^c	0.13
OM (%DM)	95.5^a	95.2^a	91.5^b	90.5^b	7.36
Ca (%DM)	0.36^a	0.38^a	1.02^b	1.14^b	0.08
P (%DM)	0.42	0.45	0.49	0.50	0.02

* 17% potassium sulfate, 45% calcium phosphate, 8% limestone 15% trace mineral salt (guaranteed analysis: 96.5% salt, 0.4% zinc, 0.16% iron, 0.12% manganese, 0.033% copper, 0.007% iodine, 0.004% cobalt), 4% Vitamin ADE premix (containing: 4,400,000 IU/kg vit.A, 1,100,000 IU/kg vit.D, 7,700 IU/kg vit.E), 11% corn, 8 mg/kg selenium.

** 40% potassium sulfate, 16 mg/kg selenium, 8% ADE Premix (as above) 22% trace mineral salt (as above), 30% corn

*** 98.8% ground corn, 1.2% Rumensin.

\+ standard error of mean.

++ Est. TDN(MFR) = ((100%-36%ash)-13%lignin) x 65% digestible.

RESULTS AND DISCUSSION

Stability tests confirmed that both sources of MFR are very stable in warm environments for several weeks. Chemical composition of these residues are presented in Table 2. Dry matter values averaged close to 22% and ranged from 17 to 29%.

TABLE 2: Nutrient composition of MFR produced commerically from beef cattle and swine wastes

Nutrient	Mean*		Range	
	Cattle	Swine	Cattle	Swine
Dry matter (%)	20.7	22.6	17.1 - 28.6	19.6 - 25.6
pH	8.3	8.4	8.0 - 8.6	8.1 - 8.6
Composition of dry matter				
Crude protein (%)	24.7	37.5	21.9 - 27.0	33.2 - 42.8
True protein (%)	14.7	18.7	9.4 - 18.4	14.6 - 22.2
NH_4-N (%)	1.54	2.89	1.16 - 2.09	2.19 - 3.48
NPN-N (%)	1.59	3.02	1.25 - 2.01	2.20 - 3.61
Ether extract (%)	8.9	11.2	5.1 - 12.3	8.5 - 16.1
ADF (%)	27.2	8.9	22.2 - 29.8	8.0 - 10.3
NDF (%)	35.7	22.3	28.7 - 39.9	19.4 - 26.2
Lignin (%)	12.6	4.3	10.1 - 15.0	2.5 - 7.8
ADF-N (%)	0.81	0.23	0.65 - 0.94	0.10 - 0.28
Gross energy (Kcal/g)	3.0	3.2	2.3 - 3.5	2.7 - 3.6
Ash (%)	36.4	36.4	30.0 - 43.1	30.4 - 42.3
Calcium (%)	6.2	5.5	5.3 - 8.9	3.5 - 6.6
Phosphorus (%)	1.6	5.1	1.3 - 2.2	2.4 - 6.5
Magnesium (%)	1.1	2.1	0.6 - 1.6	0.7 - 3.0
Potassium (%)	1.2	1.0	0.5 - 1.6	0.7 - 1.5
Manganese (ppm)	281	743	241 - 382	240 - 1147
Copper (ppm)	56	500	43 - 67	201 - 786
Zinc (ppm)	240	1356	185 - 300	481 - 2215
Iron (%)	0.38	0.44	0.11 - 0.52	0.15 - 0.94
Sodium (%)	0.55	0.27	0.33 - 0.73	0.20 - 0.31

* 12 observations per mean with samples taken every two weeks.

Cattle MFR was not attractive as a feed source simply based on chemical composition. Crude protein content was markedly lower than swine MFR. The high lignin and ADF contents would decrease digestible energy. Both MFR sources contained a large amount of ash which caused a depression in energy contents despite the relatively high levels of ether extract. Analyses for cattle MFR are similar to results of others (Prior et al., 1981; Harris et al., 1982).

Swine MFR appeared more promising as a feed source based on higher levels of crude protein and phosphorus as well as lower levels of lignin and ADF-N.

Chemical composition of diets fed during the feedlot trial is outlined in Table 1. Feedlot performance of steers fed cattle MFR is presented in Table 3 and Table 4. Animals supplemented with MFR displayed markedly depressed weight gains and required more feed per gain ($P<.01$) than animals supplemented with either urea or soybean meal. Harris et al. (1982) found a similar reduction in performance of feedlot cattle when substituting a combination of flaked corn, distillers solubles and urea with MFR at a level of 10.6% of the total dietary dry matter. Reduced performance of cattle has also been demonstrated by Johnson et al. (1981), when supplementing a diet with sewage sludge at a level of 11.6% of the total dry matter. The study concluded that sludge containing 24.9% crude protein and 45.0% ash had little value as a feed ingredient for beef cattle.

TABLE 3: Performance of growing steers fed cattle MFR over 84 days

Items	Nitrogenous feed source				
	Urea	SBM	MFR_1	MFR_2	SEM+
No. steers	20	20	20	20	
Initial wt. (kg)	294	297	292	296	4.56
Final wt. (kg)	400^a	410^b	371^b	367^b	6.61
Wt. gain (kg/day)	1.26^a	1.34^a	0.95^b	0.84^b	0.04
DM intake (kg/day)	7.45	7.58	7.39	7.22	0.13
DM intake/gain	5.94^a	5.63^a	7.97^b	8.61^b	0.23

ab means with different superscripts differ ($P<0.1$).

+ Standard error of mean.

TABLE 4: Performance of growing steers fed swine MFR

	Days on Experiment				Mean
	0-28	29-56	57-84	85-112	0-112
Liveweight gain (kg/day)					
Control	1.57	1.61	1.07	1.22	1.36
MFR	0.88	1.13	.72	.73	0.86
DM intake (kg/day)					
Control	5.96	7.22	6.84	7.21	6.81
MFR	5.51	5.90	6.20	6.94	6.14
DM intake/gain					
Control	3.80	4.48	6.39	5.91	5.15
MFR	6.26	5.22	8.61	9.51	7.40

The reduced performance was not caused by any reduction in dry matter intake, which suggests that no palatability problems existed when including MFR in diets at a rate of 14% of diet dry matter. In addition, no added health problems were experienced in cattle fed MFR versus control diets.

Digestibilities of cattle and swine MFR for dry matter, nitrogen, organic matter and gross energy were: 37 and 49%, 32 and 61%, 52 and 59%, and 44 and 52%, respectively. The reduced dry matter digestibility of the cattle MFR compared to swine MFR is partially due to its increased lignin and ADF content. The low nitrogen digestibility of cattle MFR may be largely due to its high ADF-N content. ADF-N accounts for 20.5% of the total nitrogen content. Harris et al. (1982) reported that ADF-N accounted for 27% of the nitrogen fraction of cattle MFR. Using an equation developed by Goering et al. (1972), a nitrogen digestibility of only 45-52% may be estimated. Approximately 40% of the crude protein fraction of cattle MFR was shown to be in the form of nonprotein nitrogen (NPN) which should be largely digestible. This would suggest that the true protein fraction of the cattle MFR is largely unavailable.

The poor nutrient digestibilities and low organic matter content of cattle MFR explain only in part the depressed performance observed in the feedlot trial. Further investigations (Jones et al., 1986) noted no adverse effect on rumen fermentation when feeding cattle or swine MFR. The addition of MFR to diets did not affect rumen pH, ammonia or volatile fatty acid concentrations. In addition, no adverse effects were noted on the rate of digestion of corn silage contained in nylon bags suspended in the rumen of steers.

The improved nitrogen digestibility found with swine versus cattle MFR is in part a reflection of its lower ADF-N content. Using the equation developed by Goering et al. (1972), a nitrogen digestibility of 69.1% can be estimated for swine MFR. An actual value of 60.6% was found in the present study.

Swine MFR would appear to have greater potential for use as a supplemental protein source for growing beef cattle based on its high crude protein content and improved nutrient digestibilities. Nevertheless, in the subsequent steer growth trial, supplementation with swine MFR markedly depressed weight gain and feed efficiency. Feed intake was also reduced but could not explain in full the poor performance. Performance was consistently depressed throughout the 112-day trial. Several tissues were biopsied for mineral levels in order to investigate if mineral imbalance was responsible for the poor performance. Results of tissue analyses are not yet available. Individually, the observed mineral levels should not have been sufficient to depress performance (NRC, 1980). However, diets containing MFR may depress feedlot performance as a result of interactions among various minerals present at high levels.

In summary, this research has shown that although MFR produced from cattle wastes has a crude protein content of 25%, it is an unacceptable source of supplemental crude protein for growing beef cattle. Swine MFR has a much greater crude protein content (38%) and improved nutrient digestibilities. However, it also produced very poor weight gains under practical feeding conditions. A final study is underway to determine if mineral imbalances may be the cause of such poor animal performance.

ACKNOWLEDGEMENTS

The authors express appreciation to Canviro Consultants Ltd., R. Bechtel and the late M. Selves for supplying MFR for this study and to the capable staff at the Elora Beef Research Centre for their assistance in conducting animal trials. This study was supported financially by the National Sciences and Engineering Research Council of Canada and by the Ontario Ministry of Agriculture and Food.

REFERENCES

Buchanan-Smith, J.G., S. Lee, D.N. Mowat and G.K. Macleod. 1984. Nutrient composition of methane fermentation residue. Can. J. Anim. Sci. 64:1079 (Abst.)

Goering, H.K., C.H. Gordon, R.W. Hemkin, D.R. Waldo, P.J. Van Soest and L.W. Smith. 1972. Analytical estimates of nitrogen digestibility in heat damaged forages. J. Dairy Sci. 55:1275-1280.

Harris, J.M., R.L. Shirley and A.Z. Palmer. 1982. Nutritive value of methane fermentation residue in diets fed to feedlot steers. J. Anim. Sci. 55:1293-1302.

Johnson, R.R., R. Panciera, H. Jordon and L.R. Shuyler. 1975. Nutritional, pathological and parasitological effects of feeding feedlot wastes to beef cattle. Symp. on Managing Livestock Wastes. pp. 203-205.

Jones, C.R., D.N. Mowat, J.G. Buchanan-Smith and G.K. Macleod. 1986. Methane fermentation residue as a protein supplement for beef cattle. Agric. Wastes. (in press).

MacDonald, R.D., M. Selves and R. Stickney. 1984. Selves Farms Ltd. anaerobic digester for co-generation and protein recovery. Ann. Mtg. Can. Soc. Agr. Eng. Paper No. 84-414

National Academy of Science-National Research Council. 1980. Mineral tolerance of domestic animals. NAS-NRC. Washington, D.C.

Prior, R.L. and A.G. Hashimoto. 1981. Potential for fermented cattle residue as a feed ingredient for livestock. In: Fuel Gas Production from Biomass. Vol. III. CRC Press. Florida. pp. 215-237.

ACKNOWLEDGEMENTS

[illegible]

REFERENCES

[illegible] and [illegible] Macleod. [illegible] [illegible]

[illegible] Smith. [illegible] of [illegible] [illegible]

[illegible] [illegible] value of [illegible] [illegible]

[illegible] nutritional [illegible] and [illegible] [illegible]

[illegible] Moore [illegible] Henderson. 1990. Whole-body [illegible] [illegible]

[illegible] [illegible] [illegible]

[illegible] [illegible] mineral [illegible] [illegible]

[illegible] [illegible] [illegible] Production [illegible] [illegible]

CHAETOMIUM CELLULOLYTICUM MICROBIAL BIOMASS PROTEIN EVALUATION WITH RATS, CHICKS AND PIGLETS

S.P. Touchburn and E.R. Chavez
Department of Animal Science and McGill Nutrition and Food Science Centre
Ste. Anne de Bellevue, Quebec H9X 1C0
and
M. Moo-Young
Department of Chemical Engineering
University of Waterloo
Waterloo, Ontario N2L 3G1

INTRODUCTION

Microbial biomass protein (MBP) produced by the fungus *Chaetomium cellulolyticum* has many attractions as a potential feed for animals. The large mycelia can be easily recovered by inexpensive filtration methods. Environmental pollution problems resulting from the accumulation of cellulolytic wastes (forest, pulp mill, agricultural) could be resolved by their transformation into a valuable resource for animal production. Earlier work with this organism (Leeson et al., 1984) indicated a large variability in the product composition and its acceptability by the experimental animals. Therefore, our initial goal was to determine if this fungus itself produced MBP material that was acceptable, digestible, utilizable and free of toxicity. To avoid the effect of undigested substrate in the final product these studies were confined to material produced on either glucose or molasses substrates (providing the main carbohydrate source). Their nutritive values were compared with that of the better known MBP feedstuffs, Torula yeast and Brewer's yeast and with another fungal protein, dried mushrooms. The Chaetomium MBP samples were produced at the University of Waterloo according to the method described by Moo-Young et al. (1979).

EXPERIMENT 1

Experimental Procedures

Weanling, male Sprague-Dawley rats housed in individual cages were fed a commercial diet for a three-day adaptation period. Ten rats were then fed each of the experimental diets and water *ad libitum* for a 4-week period.

The chemical analyses of the MBP samples from glucose and molasses substrates are given in Table 1. Amino acid analysis was conducted using ion exchange chromatography with ninhydrin detection on a Varian Model 5500 high pressure liquid chromatograph. The results are reported in Table 2.

TABLE 1: **Chemical analysis of different batches of single cell protein production with *(C. cellulolyticum)***

	Substrate	
	MBP-1 Glucose	MBP-2 Molasses
Chemical Analysis		
Dry Matter, %	89.58	92.55
As % D.M.		
Crude Protein, %	45.5	45.08
Ether Extract, %	1.45	1.83
Acid Detergent Fiber, %	22.57	13.05
Ash, %	8.45	10.30
Gross Energy, Kcal/Kg	4,662	4,358
Calcium, %	.06	2.59
Phosphorus, %	1.66	1.43
Sodium, %	0.80	0.61
Potassium, %	0.90	1.98
Magnesium, %	0.14	0.06
Copper, ppm	83	64
Zinc, ppm	221	182
Iron, ppm	188	637
Manganese, ppm	8	28

The compositions of the experimental diets are shown in Table 3. The test material was mixed with purified feed ingredients to provide 10 percent crude protein and the diets were fed in pelleted form. Casein supplemented with D, L-methionine served as the protein source of the control diet and a commercial rat feed was included as a reference treatment.

TABLE 2: Amino acid profile of single cell protein sample of *C. cellulolyticum* (amino acid as percent of the sample)

Amino Acid	MBP-1 Glucose	Skim Milk
Aspartate	3.08	2.28
Threonine	1.95	1.17
Serine	1.54	1.44
Glutamate	5.00	5.71
Proline	1.58	2.65
Glyine	1.79	0.56
Alanine	4.18	1.04
Cystine	n.d.	n.d.
Valine	1.99	1.73
Methionine	0.62	0.66
Isoleucine	1.68	1.34
Leucine	2.57	2.64
Tyrosine	1.20	1.17
Phenylalanine	1.18	1.14
Histidine	1.14	0.82
Lysine	2.54	2.11
Ammonia	1.07	0.66
Arginine	2.36	0.96

n.d. = not detectable amount.

TABLE 3: Composition of the experimental rat diets

Ingredient %	Casein Control Diet 2	MBP-1 Diet 3	MBP-2 Diet 4	Torula Yeast Diet 5	Brewers' Yeast Diet 6	Mushrooms Diet 7
Corn Starch	72.36	59.07	59.63	65.08	57.76	50.92
Casein	11.14	--	--	--	--	--
Protein Test Ingredient	--	24.53	23.97	18.52	25.84	32.68
Corn Oil	10.0)					
Mineral Premix	4.0)					
Vitamin Fortification Mix	2.2)			Same		
Choline-Chloride	0.2)					
Dl-Methionine	0.1)					
Chemical Analysis (%, as fed basis)						
Crude Protein	10.92	11.36	10.10	9.14	9.62	10.54
Ether Extract	10.54	11.49	11.37	11.10	11.51	11.87
AD Fibre	0.72	5.51	3.04	0.00	0.00	2.92
Ash	2.92	4.92	5.17	4.03	4.88	6.26

During the 4th week of the study the apparent dry matter and crude protein digestibilities of the diets were determined by the total collection method during 5 consecutive days. These data were also used to calculate the protein efficiency ratio (PER) of the dietary protein sources.

The data were analyzed using the General Linear Models procedure (SAS Institute Inc., 1982).

Results and Discussion

Proximate analysis of the fungal MBP samples (Table 1) indicated a very striking difference in acid detergent fibre (ADF) content (22.6 vs 13.1%) and ash content (8.5 vs 10.3%) between the glucose and the molasses-grown products. The greater ADF level of the MBP-1W sample may indicate a higher proportion of fungal cell wall (chitin) content reflecting the stage of cell maturity at harvest. The higher ash of the molasses-grown MBP-2 was due to its higher content of calcium and potassium. The crude protein contents of the two fungal MBP samples agree with values reported by Moo-Young et al. (1979) but are much higher than those reported by Leeson et al. (1984) for similar MBP products.

The amino acid profiles, after acid hydrolysis of the two fungal MBP samples (Table 2) compared favorably with that of dried skim milk.

The casein control diet showed the greatest body weight and feed consumption of all the treatments and compared well with the performance of the rats fed the commercial rat feed (Figure 1). Among the MBP products tested, Brewer's yeast gave the greatest body weight at 4 weeks of age, followed by MBP-1 and Torula yeast. MBP-2 and dried mushrooms gave the lightest body weight and lowest feed consumption. The reduced feed consumption of the rats fed the latter suggest a problem of acceptability. The MBP-2 was observed to be more dense, granular and darker in colour and to have a stronger odor.

The PER values for the different protein sources (Table 4) showed the highest value for the casein control diet. MBP-1 gave a significantly higher value than MBP-2, which may reflect the significantly lower level of protein intake in the latter although it may also reflect real differences in protein quality. The Brewer's yeast diet, while lower in PER than the control diet, showed the highest PER of the MBP samples, being significantly greater than that for MBP-1. The PER for Torula yeast was also greater that MBP-1. The very low PER for MBP-2 and for dried mushrooms which were similar, may be at least partly due to the very low intake of these two diets.

Data on crude protein digestibility are shown in Figure 2. Casein showed the highest value (93%), followed by Torula yeast (82%). The MBP-1 sample was superior to MBP-2, the former averaged 79% higher and was similar to the value for Brewer's yeast (77%) and significantly higher than that for MBP-2 (68%). Mushroom protein showed the lowest digestibility (65%). However, the low intake of the MBP-2 and mushroom diets would lead to a considerable underestimation of the digestibility values for these two protein sources.

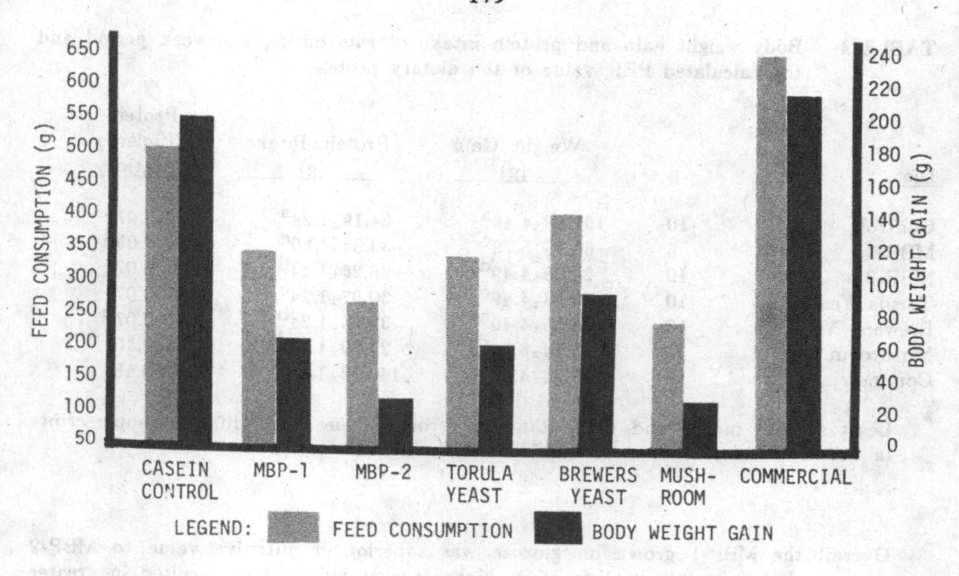

FIGURE 1: Rats: four week feed consumption and body weight gain.

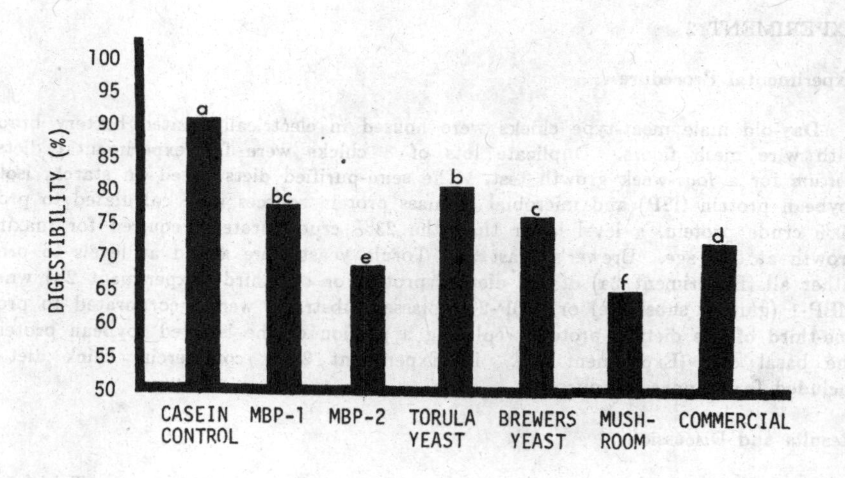

FIGURE 2: Apparent digestibility of crude protein of rat diets at four weeks

TABLE 4: **Body weight gain and protein intake of rats during a 4-week period and the calculated PER value of the dietary protein**

Diet	n	Weight Gain (g)*	Protein Intake (g)*	Protein Efficiency Ratio*
Control	10	191.88±4.49[a]	54.19±1.24[a]	3.54±0.07[a]
MBP-1	6	65.95±5.79[c]	39.34±1.60[c]	1.66±0.09[d]
MBP-2	10	26.08±4.49[d]	26.26±1.24[d]	0.99±0.07[e]
Torula Yeast	10	60.56±4.49[c]	30.97±1.24[c]	1.95±0.07[c]
Brewers' Yeast	10	93.25±4.49[b]	39.25±1.24[b]	2.37±0.07[b]
Mushroom	10	25.74±6.34[d]	24.29±1.76[d]	1.06±0.10[e]
Commercial	10	215.72±4.39	142.52±1.36	1.52±0.18

* Least squares means and their standard errors. Values with different superscripts in the same column indicate significant differences ($P<.05$).

Overall the MBP-1 grown on glucose was superior in nutritive value to MBP-2 grown on molasses, mainly because of its higher acceptability which resulted in greater intake. In contrast, the intake and protein digestibility values of MBP-1 were similar to those of Brewer's yeast. Thus, the superior growth of rats fed the Brewer's yeast diet must be attributed to its superiority in either amino acid pattern or availability which was also reflected in a superior PER.

EXPERIMENT 2

Experimental Procedure

Day-old male meat-type chicks were housed in electrically-heated battery brooders with wire mesh floors. Duplicate lots of 8 chicks were fed experimental diets *ad libitum* for a four-week growth test. The semi-purified diets based on starch, isolated soybean protein (ISP) and microbial biomass protein sources were calculated to provide 20% crude protein, a level lower than the 23% crude protein required for maximum growth at this age. Brewer's yeast and Torula yeast were added at levels to provide either all (Experiment 2a) of the dietary protein or one-third (Experiment 2b) whereas MBP-1 (glucose substrate) or MBP-2 (molasses substrate) were incorporated to provide one-third of the dietary protein, replacing a portion of the isolated soybean protein of the basal diet (Experiment 2b). In Experiment 2a a commercial chick diet was included for comparison purposes.

Results and Discussion

Performance records of the chicks to 4 weeks of age are shown in Table 5. In Experiment 2a the ISP diet yielded a significantly greater body weight gain than either Brewers' yeast of Torula yeast diets. In Experiment 2b the combination of ISP plus MBP-1 resulted in a significantly greater weight gain than when ISP was incorporated as the sole source of protein. No significant differences in body weight were observed among diets containing ISP alone or in combination with MBP-2, Brewers' yeast or Torula yeast. As in the preceding rat growth trial, the MBP-1 appeared to be of better quality than MBP-2.

TABLE 5: Performance* of chicks over a 4-week experimental period

Diet	Weight Gain (g)	Feed Consumption (g)	Feed Conversion Feed/Gain
Experiment 2a			
Control	313.88[b]	624.37[b]	2.11[a]
Brewers' Yeast	210.56[c]	478.88[c]	2.63[a]
Torula Yeast	203.70[c]	439.41[d]	2.39[a]
Commercial Feed	1043.62[a]	1556.74[a]	1.50[b]
S.D.	63.38	45.33	0.83
Experiment 2b			
Control	245.00[b]	503.18[e]	2.38
MBP-1 + ISP	341.88[a]	700.50[a]	2.17
MBP-2 + ISP	299.27[ab]	648.18[b]	2.31
Brewers Yeast + ISP	288.47[ab]	566.58[c]	2.05
Torula Yeast + ISP	280.81[b]	544.68[d]	2.11
S.D.	81.39	26.18	0.70

* Least squares means. Values with different superscripts in the same column indicate significant difference ($P < .05$).
S.D.Standard deviation. Sixteen observations per dietary treatment.

Feed consumption data, also shown in Table 5, were reflected in the body weight gains with consumption and gain of the commercial diet being much greater than that of the experimental diets as would be expected. Feed efficiency showed no significant differences among dietary treatments except for the lower efficiency of the commercial diet.

The Chaetomium MBP samples in combination with ISP appeared to be well accepted and resulted in growth performance better than or at least equal to that of the better known MBP materials, Brewers' yeast and Torula yeast.

EXPERIMENT 3

Experimental Procedures

Ten purebred Landrace piglets weaned at two weeks of age were used. Six female and four castrated male piglets were first set in five pairs of lettermates of the same sex and similar weight, then one of each pair was assigned to one of the two experimental groups. They were housed and fed individually in wire-mesh floor cages with free access to feed and water and received a standard commercial pre-starter pig ration for an adaption period of four days prior to the experiment.

Results and Discussion

Performance records of the piglets are shown in Table 6. No significant differences were observed between the control and MBP groups with respect to body weight or body weight gain when initial weight was used as a covariate at any time during the experimental period. During the first week of the experiment the control group consumed more feed ($P<.05$) than the test protein group. However, no significant differences were observed between these two experimental groups in feed consumption during the second week, or in the total for the 19-day experimental period.

Although the data are not presented, mean blood parameters of individual samples taken at day 19 of the trial have shown that control and MBP groups had similar blood hematocrit values and contents of hemoglobin, Cu, Zn, Fe and Mn. Plasma levels of protein, glucose, urea and creatine were likewise similar for the two dietary groups.

Thus, Chaetomium MBP included at approximately 14% in the diet of weanling pigs could replace the equivalent amount of protein from soybean meal without affecting the performance of these fast growing animals. This represents the first report on the biological value of Chaetomium MBP for pigs.

EXPERIMENT 4

The objectives of this study were to determine the effect of the drying method on the fungal MBP quality and to evaluate the animal response to amino acid supplementation of fungal MBP containing diets.

Experimental Procedure

A Chaetomium MBP batch culture (PP101) produced on glucose substrate as previously described was divided in half. One half was dried in a forced air oven at 60 °C (MBP-OD), the other half was frozen at -25 °C, then dried in a freeze-drier (MBP-FD). Both samples ultimately contained more than 90% dry matter.

TABLE 6: Performance of weanling pigs during a 19-day experimental period

Parameter	Diet		S.D.[2]
	Control	Chaetomium MBP	
Body weight (g)			
Initial	4981	4815	830
Final	11300	11740	1538
Body wt. gain[1] (g)			
1st wk	1256	1213	224
2nd wk	3000	3300	550
19 d	6285	6960	1059
Feed consumption [1] (g)			
1st wk	1400[a]	1242[b]	107
2nd wk	2715	2978	442
19 d	6620	7121	789
Feed conversion[1,3]			
1st wk	1.11	1.02	0.08
2nd wk	0.90	0.90	0.13
19 d	1.05	1.02	0.12

1 Least squares means corrected for initial weight. Values with different letters in the same row indicate significant differences (P<.05).
2 Standard deviation. Five observations per dietary treatment.
3 Grams of feed to gain one gram of body weight.

Weanling male Sprague-Dawley rats housed in individual cages were fed a commercial rat diet during a three-day adaptation period. They were then assigned to groups of 10 rats per dietary treatment and fed the experimental diets for a 2-week period.

The diets prepared as in Experiment 1 included: 1, a control containing casein as the sole source of protein, three experimental basal diets without amino acid supplementation (2, 5 and 8) as follows: diet 2 included isolated soybean protein (ISP) as the sole source of protein while diets 5 and 8 included one half of the protein supplied by ISP and the other one-half provided by MBP-OD and MBP-FD respectively. The three basal diets were also tested after being supplemented with synthetic amino acids, 0.4% DL-methionine (diets 3, 6 and 9) or with 0.4% DL-methionine plus 0.15% L-lysine (diets 4, 7 and 10).

Results and Discussion

Significant differences were observed among the experimental groups in body weight gain, feed consumption and feed/gain ratios (Table 7). The control-casein diet gave a significantly greater body weight than the ISP diet or ISP blended with MBP product obtained after oven- or freeze-drying treatment. No significant difference was observed in the body weight gain between the ISP basal group and the basal containing ISP blended with the FD sample. However, the basal containing the OD sample gave a significantly lower body weight gain than the other two basal diets. Thus, it would appear that the FD sample had a higher protein value than the OD sample, indicating that the freeze-drying process resulted in a better MBP product that the oven-drying process. This is in disagreement with previous results which indicated that the oven-drying and freeze-drying methods do not influence the true protein content as amino acids in this kind of fungal MBP product (Leeson et al., 1984). This conclusion on the effect of drying on product quality has to be considered as tentative, however, because of the significant differences in feed consumption.

TABLE 7: Mean performance of the rats receiving different experimental diets during a 2 week period

Diet	Body weight gain (g)	Feed consumption (g)	Feed/gain
1. Casein - control	78.1^{ab}[2]	178.4^{bc}	2.30^{e}
2. Isolated soybean protein (ISP)	56.7^{d}	177.2^{bc}	3.16^{c}
3. ISP+DL-Met	67.6^{c}	174.1^{c}	2.59^{de}
4. ISP+DL-Met+L-Lys	69.4^{bc}	178.3^{bc}	2.59^{de}
5. PP101-OD	32.1^{d}	131.1^{d}	4.16^{a}
6. PP101-OD+Met	78.1^{ab}	209.5^{a}	2.72^{d}
7. PP101-OD+Met+Lys	80.9^{a}	210.6^{a}	2.66^{d}
8. PP101-FD	49.3^{d}	167.2^{c}	3.44^{b}
9. PP101-FD+Met	79.3^{ab}	193.6^{ab}	2.45^{de}
10. PP101-FD+Met+Lys	84.1^{a}	194.7^{ab}	2.33^{e}
S.D.[1]	10.8	18.1	0.30

1 Standard deviation. Ten observations per dietary treatment.
2 Values with different letters in the same column indicate significant differences ($P < .05$)

All three basal diets (ISP, MBP-OD and MBP-FD) showed significant responses to supplementation with D, L-methionine. Further supplementation with L-lysine resulted in slight improvements in body weight gain in each case but the differences were not significant. The double supplemented MBP diets, however, showed a weight gain that was significantly greater than the ISP diet similarly supplemented.

Feed/gain ratio as an indicator of feed and dietary protein utilization efficiencies indicated that the casein in the control diet gave the best result although the ISP basal and the MBP-FD basal diets, when singly or doubly supplemented with amino acid(s), resulted in similar feed/gain ratios. The MBP-OD basal, when supplemented with DL-methionine or DL-methionine plus L-lysine, resulted in a good body weight gain of the rats but feed/gain ratios were higher, indicating a lower feed and dietary protein utilization efficiency that proved to be significantly poorer than the control group or the fully supplemented MBP-FD group.

Since the incorporation with appropriate supplementation of the MBP product in a complete diet resulted in similar performance regardless of the method of drying, the selection of this drying method can be dependent on the cost of the process. Finally, the tremendous improvement in animal performance in response to amino acid supplementation of this fungal MBP product provides good grounds for optimism for the future use of this non-conventional protein source in the animal production system.

In general the positive response to Chaetomium MBP by rats, chicks and piglets indicates that this new source of protein will likely be effectively used by most domestic animals.

ACKNOWLEDGEMENTS

This work was supported by a Strategic Grant from the Natural Sciences and Engineering Research Council of Canada.

REFERENCES

Leeson, S., Summers, J.D. and Lee, B.D. 1984. Nutritive value of single cell protein produced by *Chaetomium cellulolyticum* grown on maize stover and pulpmill sludge. Anim. Feed Sci. Technol., 11: 211-219.

Moo-Young, M., Daugulus, A.J., Chahal, D.S. and MacDonald D.G. 1979. The Waterloo Process for SCP production from waste biomass. Process Biochem. 14: 38-40.

SAS Institute Inc. 1982. SAS User's Guide: Statistics. 1982 ed. SAS Institute Inc., Cary, N.C.